Power
to the
People

MARTIN GREEN

Power to the People

Sunlight To Electricity Using Solar Cells

UNSW
PRESS

A UNSW Press book

Published by
University of New South Wales Press Ltd
University of New South Wales
Sydney 2052 Australia
www.unswpress.com.au

© Martin A. Green
First published 2000

National Library of Australia
Cataloguing-in-Publication entry:

Green, M.A.
Power to the people: sunlight to electricity using solar cells.

Bibliography.
Includes index.
ISBN 0 86840 554 X.

1. Solar cells. 2. Photovoltaic power generation. I. Title.

621.31244

Printer Southwood Press

CONTENTS

PREFACE

Most people who have heard of photovoltaics and solar cells have heard about them through solar car racing. Most are fascinated by their elegance and their sheer magic — the way the cells silently and effortlessly change sunlight to electricity.

Many more will hear about solar cells over the next decade. Millions will use them on their rooftops to supply electricity to their homes. The first being affected are homes at two extremes — those in the suburbs of the wealthiest countries of the world and those in the most isolated rural areas of the poorest.

This book is based loosely on material presented in public lectures with my colleague, Stuart Wenham, following the award of the 1999 Australia Prize for 'Outstanding Achievement in Science and Technology', based on our work on improving and commercialising solar cells. It takes the reader step-by-step through the way solar cells work and the improvements that are driving down their costs. The book then focuses on present and future uses of the cells, particularly three key ones for the coming decade — on urban residential rooftops, architecturally in buildings, and in the developing world.

I would like to thank the many who have stimulated my interest in photovoltaics over the years and particularly Jenny Hansen who, amongst other things, computerised many of the original drawings in the text. I also thank my family — Judy, Brie and Morgan — for their tolerance of my ongoing immersion in projects such as this book.

Martin A. Green

1

INTRODUCTION

Sunlight and solar cells

We all know that huge amounts of fuel are being burnt to meet the world's insatiable energy demands. We also know how damaging this is to the environment. It therefore comes as a surprise to discover that in only a few days, the earth receives more energy from the sun than from all the fuel burnt over the whole of human history. Three weeks of sunshine offsets all known fossil fuel reserves.

The sun already provides almost all the energy needed to support life as we know it. The challenge for a sustainable future is to tap into a tiny fraction of this energy to supply the relatively modest demands of human activity. Probably the most elegant way known of doing this is to convert it straight into electricity using solar cells, or 'photovoltaic' cells as they are also known ('photo' refers to 'light' and 'voltaic', 'electricity'). This is quite a new approach, not yet fully mature, based on some of the major achievements in science and technology of the 20th century — including the development of quantum mechanics, microelectronics and the conquest of space.

In spite of this high-technology background, solar cells are surprisingly easy to use. Placed in sunlight, they just soak up the sun's energy and change it into electricity, on tap at their terminals. These cells, however, seem destined to change the way we think about energy and its production.

How the cells work

Figure 1 shows a typical present-day solar cell. The cell is made from a thin wafer of silicon (the same material used in microelectronics), approximately 10 x 10 cm in size — about the size of a compact disc — and only a fraction of a millimetre thick. A patterned metal layer partly covers the side exposed to sunlight making electrical contact to it, with a second metal contact covering most of the rear surface. When sunlight falls on the cell, an electrical output is generated between these contacts, allowing the cells to be used as a battery that can, in principle, last as long as the sun shines.

Solar cells are seldom sold individually. Generally, 36 cells are connected within a weatherproof package known as a solar or photovoltaic 'module' (Figure 2). A sheet of strengthened glass forms the top surface of this module to protect the cells from the harsh outdoor environment where they operate. Several manufacturers warrant modules for up to 25 years, a warranty period matched by few other manufactured goods (saucepans are one of the few products with a similar warranty period).

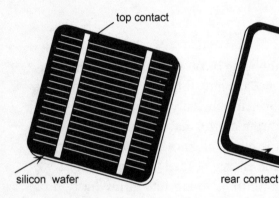

Figure 1
A silicon solar cell shown from the top and rear. Sunlight enters from the top and is absorbed within the silicon wafer that forms the body of the cell, creating an electrical output between the top and rear contacts.

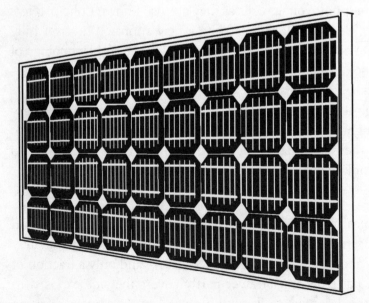

Figure 2
A standard solar cell module showing 36 cells connected together into a weatherproof package with a strengthened glass sheet forming its top surface.

When in sunlight, solar modules can power electrical loads in nearly the same way as a car battery. In the past, their main use has been for generating small amounts of electricity in areas where there is no other electricity available, such as remote rural areas in Australia. With decreasing solar cell costs and the urgent need to find better ways of supplying the world's energy, modules are being used in rapidly increasing numbers in urban areas — particularly on the family home. Further into the future, as the industry grows and prices drop further, solar cells will be used side-by-side with conventional large-scale power plants as shown in Figure 3. In the very long term, past 2050, almost all the world's energy could be generated by these cells.

Why not right now?

Solar cells are safe, clean, quiet, durable and reliable and can be installed almost anywhere they can 'see' the sun — on the roofs of homes or motor vehicles, integrated into building facades, on snowy mountain peaks, or in desert wastelands.

Where's the catch? Why aren't we using more right now? In the past, the problem has been the high costs of the cells. The electricity they produce has been much more expensive than conventional methods of generating electricity. Of course the real cost difference would be less if environmental and social costs were taken into account.

All this is changing. As shown in Figure 4, the costs of the cells have reduced markedly as the number made has increased. The good news is that this trend will continue, driven both by increased manufacturing volume and new, lower-cost solar cell technologies.

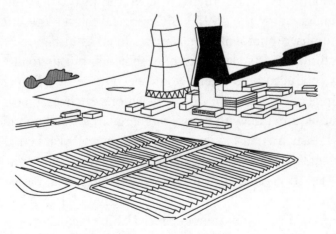

Figure 3
A sketch of the Rancho Secco power station in California with a large photovoltaic field (foreground) installed next to a nuclear power plant (the latter is being decommissioned after increasing safety concerns prompted a public referendum in 1989, sealing its fate).

Solar cells have now become 'affordable' to many more people. Anyone who can afford to buy a home in the western world can afford to put solar cells on the roof to generate most of the home's electricity needs — or even to sell surplus electricity back to the power company.

This realisation is driving a massive growth in their use, with governments in Japan, USA and Europe battling to see who can be the first to have one million homes powered by solar cells (all have announced the year 2010 as the target date for reaching this goal).

These initiatives are increasing confidence within the industry as well as increasing market size. Both are expected to reduce manufacturing costs substantially. Within a decade the use of solar cells on private urban homes may drive costs down to the level where solar cells are more broadly competitive — even using standard economics that ignore the environmental and social costs of the conventional approaches.

Within two to three decades, centralised power stations like the one shown in Figure 3 are also likely to become competitive. Eventually, solar cells could provide the cheapest of all known ways of generating electricity.

Future scenarios

In a study exploring various energy supply futures, including 'clean' coal and 'safe' nuclear power, Bent Sorenson of Rothskilde University, Denmark, has calculated that covering just 1 per cent of the area of urban centres with photovoltaics, using roofs and building facades, and 0.01 per cent of farmland (about 25 per cent of farmhouse roofs) would generate more electricity than that produced worldwide in 1998. Alternatively, covering just 1 per cent of the land classified as 'marginal' by the 1997 US Geological Survey (deserts and scrubland) with centralised photovoltaic power stations, as in Figure 3, would supply not only all the world's electricity requirements, but its total energy needs (the electricity produced by the cells would need to be stored in chemicals, for example, hydrogen or methanol, for applications such as motor vehicles or aeroplanes that require fuel with plenty of energy in a small volume).

How feasible is such a transition in energy use over a 50-year period? Many think it is feasible, including some of the key players in the current energy industry. For example, Cör Herkströter, the chairman of Royal Dutch/Shell, has speculated that half the company's business will be in renewable energy, such as solar and wind, by 2050, with the company recently announcing a US$0.5 billion, five-year program to increase the company's involvement with renewables, particularly solar cells. Another oil giant, British Petroleum, has made similar announcements, with the head of its solar division stating: 'One day this industry (solar) will be as big as oil.'

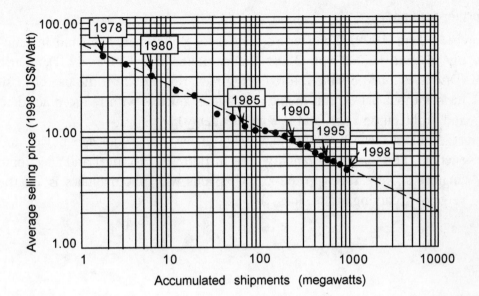

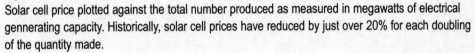

Figure 4
Solar cell price plotted against the total number produced as measured in megawatts of electrical
gennerating capacity. Historically, solar cell prices have reduced by just over 20% for each doubling
of the quantity made.

In the nearer term, one thing that solar cells can do that is difficult with conventional approaches is to raise the standard of living for some of the world's
poorest — those living in rural areas in the developing world. Some two billion
people worldwide do not have access to electricity because it is too expensive to
supply by conventional means — long power lines carrying the small amounts
of electricity able to be afforded. Solar cells can do this job much better.

***

The early part of this book explains how solar cells almost magically convert
sunlight into electricity — a legacy of Albert Einstein's fundamental contributions to the theory of light and the 'photoelectric' effect. The book then explores
the chequered history of solar cell development since the early 1940s, when the
first silicon cells were discovered accidentally, and introduces the 'thin-film'
solar cells that make the future look so promising.

The second part of the book describes activities around the world promoting the residential use of photovoltaics including how home owners can
become involved, and why they would want to. After describing some of the
visually exciting ways cells are being used in architectural projects, it describes
how the cells might also meet the needs of those in developing countries.

Further reading

Sorenson, Bent (1999), 'Long term scenarios for global energy demand and supply', *Report TEKST,* vol. 359, January, Rothskilde University, Denmark. (Discusses four possible future energy scenarios based on the use of fossil fuels without net carbon dioxide emissions, nuclear with reduced accident and proliferation risks, and renewable energy.)

Street, William (1999), 'Technology 1999 analysis and forecast: power and energy', *IEEE Spectrum*, January, The Institute of Electrical and Electronic Engineers, New York, pp. 62–67. (Explains why photovoltaics is 'on the verge of becoming big business'.)

2

QUANTUM PHYSICS, SEMICONDUCTORS AND SOLAR CELLS

To understand how a solar cell produces electricity from sunlight, we must briefly touch base with some of the most exciting developments in 20th century science and technology. We need to go back to the beginning of that century to the birth of quantum physics. We then jump 50 years to the early days of microelectronics.

Einstein's light quanta

Some of the greatest minds in history have tried to understand light. Isaac Newton (1642–1727), one of the 'giants' of science, made great leaps forward in the study of light and its properties. He thought of light as a stream of tiny particles like miniature billiard balls. Experiments in the 18th and 19th centuries showed that this had to be wrong — light had to act as a wave, like ripples on a pond. This explained what are known as 'interference' effects — the band of colours often seen on the surface of soap bubbles or in oil-slicks on the road, and even the colours of the rainbow.

At the beginning of the 20th century, physicists were terribly confused because they couldn't explain the properties of light coming from hot bodies such as the sun. Max Planck (1858–1947) showed he could explain nature's actual behaviour if he assumed that changes in energy within the hot body could occur only in small steps known as 'quanta'. This clue triggered a revolution in 20th century physics.

Albert Einstein (1879–1955) is very well known for his work on relativity. However, he made outstanding contributions to several other areas of physics, including the physics of quanta or 'quantum physics'. His somewhat overdue Nobel Prize in 1921 was awarded for 'his services to theoretical physics and especially for his discovery of the law of the photoelectric effect', which involved Einstein's key contribution to quantum physics. Einstein's contributions in this area are described in the introduction to his 1905 paper on light quanta. Abandoning the classical picture of light as a wave, Einstein proposed that the energy of light was not spread continuously in space but (in English translation) 'consists of a finite number of energy quanta localised at points of space that move without dividing, and can be absorbed or generated only as complete units' — back to the corpuscular ideas of Isaac Newton.

Einstein pointed out that the well-known wavelike properties of light did not contradict this corpuscular interpretation. In its effects, light behaves neither as a wave nor a billiard ball, but as something foreign to everyday experience. Following Einstein, we can think of light from the sun coming in tiny packets of energy as shown schematically in Figure 5. These light quanta are now known as 'photons'.

The eye cannot see all the photons coming from the sun, but only visible light that ranges through the seven colours of the rainbow, from red through to violet. The size of each packet (in terms of the energy it contains) is about twice as large for light at the violet end of the rainbow's spectrum as at the red end. The packet size is even larger for ultraviolet light, which is the reason it does

Figure 5
A quantum picture of light from the sun. Light comes from the sun in packets of energy known as photons visible to the eye as colours.

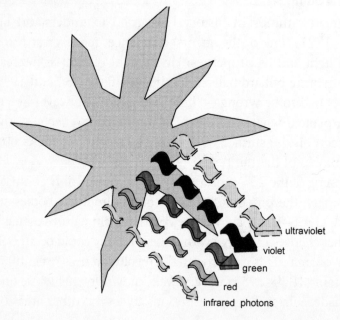

give us sunburn. Beyond the red light there is low-energy infrared light. Again, the eye cannot see this although devices like 'heat-sensing cameras' can. Although the packet size is largest for the ultraviolet photons, there are, fortunately for us, fewer of them. Most total energy from the sun (packet size multiplied by the number of packets) actually comes at the red and green wavelengths to which our eyes are most sensitive. This is undoubtedly not a complete coincidence — the result of evolution or grand design, or both, depending on your point of view.

The evidence that Einstein called upon to support his light quanta ideas came from the experiments of other scientists studying the photoelectric effect, involving the interaction of light with metal conductors. The closely-related 'photovoltaic' effect, on which solar cells are based, involves light interaction with materials known as 'semiconductors'. These materials have properties in between those of metals which are good conductors of electricity, and very poor conductors, known as insulators.

Semiconductors

Silicon is the most common semiconductor, underpinning the microelectronics industry, the computer revolution, the information age, and the other rapidly growing areas revolutionised by modern electronics. Silicon has an 'atomic number' of 14, which means that an isolated atom of silicon consists of 14 electrons (negatively charged particles) surrounding a dense central nucleus (positively charged core), like a miniature solar system. Ten of these 14 electrons are very tightly bound to this core and are not of any further interest, at least not for solar cells.

The four electrons left remaining determine how silicon atoms arrange themselves when they form solid silicon material. Solid silicon for solar cells is made by extracting silicon from sand (a compound of silicon and oxygen), melting and then slowly cooling it. Silicon freezes with the atoms doing their best to arrange themselves in a very particular pattern. Each silicon atom tries to link with four neighbouring atoms as in Figure 6. The 'glue' bonding the atoms together is two shared electrons, one from each atom. Remembering that each silicon atom has four electrons that are not tightly bound, everything works out neatly if each silicon atom is surrounded by exactly four other atoms.

As electricity is just the flow of electrons, silicon is a poor conductor of electricity when all electrons are constrained in bonds as in Figure 6 — so acts as an insulator. However, these bonds can be broken if sufficiently jolted — for example, by an energetic photon from the sun. Once released from a bond (as shown in Figure 7), the electron can move through the silicon and contribute to electrical current flow. Silicon with broken bonds acts like a conductor.

This gives a clue as to why silicon is known as a semiconductor. Sometimes it acts like an insulator, and sometimes as a conductor. We are now close to understanding how a solar cell works.

The electrons released from bonds are able to move through the semiconductor. More surprisingly, the broken bonds themselves can move. This is because it is very easy for an electron in a neighbouring bond to jump into the vacant spot left by the broken bond (see Figure 7). This jump restores the originally broken bond but leaves a new broken bond behind.

In this way, the broken bond can move through the silicon. To visualise this motion, the broken bond can be thought of as a particle (called a 'hole'), something like a bubble. Just as two negatives make a positive, the hole has an electrical charge opposite to that of the released electron. When a photon breaks a bond in silicon, a negatively charged electron and a positively charged hole are created, known as an 'electron-hole pair'.

Silicon's versatile properties

The properties of silicon can be altered in other ways than by shining light on it — by adding small amounts of impurities. (This is one reason why it is very important in microelectronics.) For example, if a small amount of phosphorus is added to melted silicon, the solidified silicon will contain phosphorus atoms in some positions where silicon would normally be, as shown in Figure 8(a).

Phosphorus has five electrons that are loosely bound to its central core. Four of these are used in the bonds between neighbouring silicon atoms, but the fifth one is at a bit of a loose end. It is only weakly bound to the original phosphorus atom and can be very easily torn away. Once separated, it acts very much the same as an electron released by light absorption.

If a different impurity such as boron, with only three electrons loosely bound to its core, is introduced in the same way, full bonds will be formed only with three of the neighbouring silicon atoms, as shown in Figure 8(b). Adding boron, or 'doping' with boron, is a good way of introducing broken bonds or holes into the semiconductor material.

Silicon doped with phosphorus is a reasonably good conductor because it has plenty of unbound or free electrons. Since it has plenty of these negative charge carriers, it is known as negative-type or n-type material. Silicon doped with boron is also a reasonably good conductor because it has plenty of holes, positive charge carriers. Not surprisingly, such material is known as positive-type or p-type material.

One of the most important devices in all of microelectronics is formed by a junction between p-type and n-type material. In fact, such p-n junctions can be regarded as the basic building blocks of microelectronics and one of the most

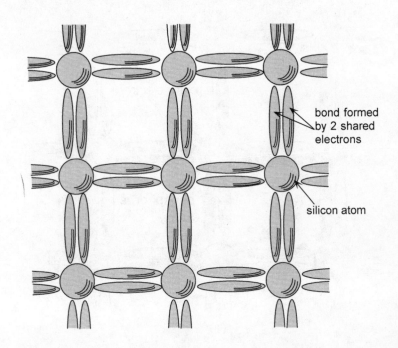

Figure 6
Sketch of how silicon atoms attempt to arrange themselves on cooling from a melt. Each silicon atom attaches to four neighbouring atoms (the actual arrangement is more interesting, involving the third dimension, but is more difficult to visualise and to draw).

Figure 7
Silicon with one electron released from a bond, by, for example, an energetic photon. The released electron is free to move through the semiconductor — so is the broken bond.

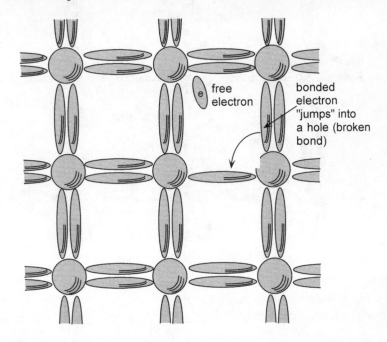

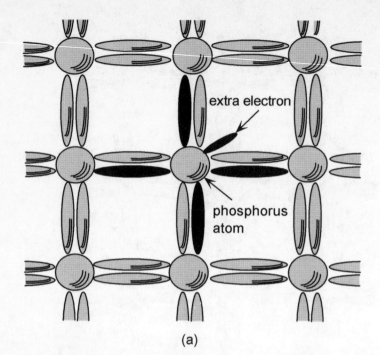

(a)

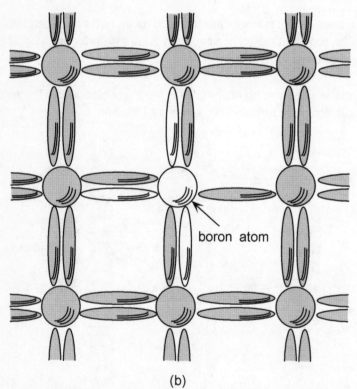

(b)

Figure 8
(a) Silicon with small amounts of phosphorus added.
(b) Silicon with small amounts of boron.

important inventions in human history. Such junctions are also key elements of solar cells.

Grand synthesis

Based on what we now know, we are in a better position to appreciate what goes on inside a solar cell. The grand synthesis is shown in Figure 9. Photons in sunlight enter into the silicon through the spaces between the top metal contact. Once in the silicon, the more energetic photons are absorbed by giving their energy to electrons originally constrained in the bonds holding the silicon atoms together. This releases electrical charge carriers, the electrons and holes, within the silicon material.

The p-n junction is the final bit of technology required to complete the picture. Something is needed to encourage all the released electrons to move off in the same direction and all the holes to move in the opposite direction.

The released electrons can flow more easily in the region where there are plenty of them, that is, the n-type side of the device. The holes (broken bonds) flow more easily in the p-type region. Although the details are subtle, the end result is that this asymmetry causes a directional flow of electrons, released by the light, from the p-type to the n-type side of the p-n junction and an opposite flow of holes. If an electrical load is connected between the n-type and p-type regions, the flow of electrons continues around through the load back to the p-type side of the device, where each electron will mate up with a hole. The bonds will be restored and the electrical circuit completed.

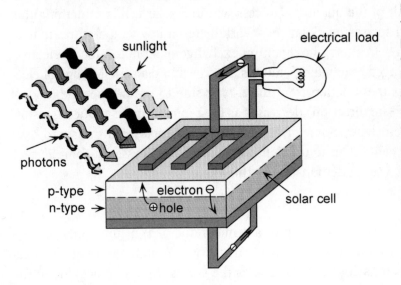

Figure 9
The photons in sunlight release electrons from the silicon bonds, creating mobile electrons and holes. The p-n junction causes these to go in opposite directions. The electrons flow through the external load and meet up with holes on their return.

The important points

The previous sections have taken you on a whirlwind tour of 20th century physics and microelectronics. Nothing important to solar cell operation has been left out. The main point to remember is that solar cells operate as 'quantum' devices. Each photon in sunlight, if energetic enough, can cause one electron to flow in the external circuit connected between the cell contacts. Photons in sunlight can be exchanged for such electrons ideally on a one-to-one basis.

This 'one-to-one' conversion is very important to the efficiency of the overall process (the efficiency is the amount of electricity the solar cell produces for a given amount of sunlight striking the cell surface). A blue photon with twice as much energy as a red photon ideally produces the same result as the red one — both produce a single electron flowing through the electrical load. The energy of the blue photon clearly is not used very effectively. Largely as a result of this effect, the efficiency of a standard solar cell is limited to about 33 per cent, no matter how well it performs. Only about one-third of the energy in the incident sunlight has a chance of being converted to electrical energy with two-thirds guaranteed to be wasted. The best commercial solar cells are only about half as efficient as the ideal case, with efficiency values generally in the 10–18 per cent range.

This is not as bad as it might appear. Much of the sunlight reaching the earth would be wasted if not converted. The energy conversion efficiency of a solar cell is a little bit different, therefore, than the efficiency of generating electricity by burning fossil fuel, since it is better to leave as much fuel as possible in the ground, given the environmental consequences of burning it.

Another result of the quantum process within a solar cell is something that most people find surprising. Solar cells work better at low temperatures than at high temperatures. In short, the quantum exchange of a photon for an electron continues unchanged at low temperatures. However, a parasitic leakage of electrons back across the junction from the n-type side to the p-type side (and of holes in the opposite direction) decreases as the temperature is reduced, resulting in increased efficiency. Some of the very best efficiencies ever measured have come from solar cells taken to the South Pole as an unnecessarily complicated, but adventurous, way of demonstrating this effect.

***

In brief, a solar cell, consisting of a semiconductor p-n junction, converts photons emitted by the sun to electrons that flow through anything electrical connected between its contacts. As long as the sun is shining, it acts much the same as a standard chemical battery, such as in a radio or car.

Further reading

Green, M.A. (2000), *Solar Cells: Operating Principles, Technology and System Applications* (2nd edn), Centre for Photovoltaic Engineering, University of NSW, Sydney. (This introductory textbook, available from the author, has been widely used in both English and foreign language versions and would be suitable for the more mathematically orientated.)

Zweibel, K. and Hersch, P. (1984), *Basic Photovoltaic Principles and Methods*, Van Nostrand Reinhold, New York.

3

THE STORY SO FAR:
A HISTORY OF
PHOTOVOLTAICS

Accidental discovery

Although there had been rudimentary solar cells made from other materials as early as 1839, the first silicon solar cell was made entirely by accident in early 1940 when Russell Ohl, a researcher at Bell Telephone Laboratories in New Jersey, made an important discovery. When he shone a flashlight onto a piece of silicon he was studying, the needle of the voltmeter connected across it jumped to a surprisingly large reading.

Investigating further, he and his colleagues found that the silicon had different properties in different areas which they called positive-type (p-type) and negative-type (n-type), depending on which side gave the more positive voltage. These terms are still used. We now know that, due to the way his silicon samples were prepared, they had different amounts of boron and phosphorus in these different areas.

This led to the discovery of a junction between these two types, the p-n junction (as explained in Chapter 2). A rapid improvement in the understanding of the key properties of semiconductors followed, leading to the birth of the microelectronics industry.

The semiconductor electronics industry evolved very rapidly during the 1950s as did the design of silicon solar cells after the demonstration of the first transistor also at Bell Laboratories in 1948. William Shockley, John Bardeen and Walter Brattain were awarded the Nobel Prize in 1956 for this breakthrough, which has led to the pervasiveness of electronics.

The first efficient solar cell, again made at Bell Laboratories, was reported in 1954, attracting front page headlines in the *New York Times*: 'Vast power of the sun is tapped by battery using sand ingredient' (26 April), and the *Wall Street Journal*. Rapid progress in techniques for preparing silicon material and for making p-n junctions in a controlled way gave massive improvements in solar cell performance during the 1950s. By the end of the decade, solar cells could convert about 14 per cent of the available sunlight into electricity.

The early cells generated enormous excitement, stimulated by the possible new uses of this technology, however they were just too expensive at that time for any but very specialised applications. One such application was soon found during the space race between the super powers.

Space program

In 1958, the first solar cells were launched into space on the US satellite 'Vanguard I'. This tiny satellite was equipped with a small solar cell module powering a radio transmitter. The cells worked so well that the transmitter kept beaming out radio signals over the next several years, clogging up the 'airwaves' in space.

It was clear that solar cells were ideally suited for space use. Throughout the 1960s, the cells were used as the power source on an increasing number of satellites. When satellites routinely started being used to relay radio, television and telephone signals from one side of the globe to the other, the future for the space solar cell manufacturing industry was secure. A vibrant industry grew making cells for spacecraft. More recently, the massive increase in satellite numbers to satisfy demand for mobile telephones and high speed internet linkages has further stimulated this space cell industry. Over 1000 solar-powered satellites are likely to be launched over the coming decade.

Oil crisis

The Middle East oil embargoes of the early 1970s emphasised the western world's dependence on oil, and on fossil fuels in general. The embargoes caused petrol shortages in many parts of the world, leading to an urgent search for alternatives to oil.

The solar cell industry, mainly involved in supplying cells for spacecraft, responded by preparing plans for the development of cheap solar cells for land use. These plans called for an enormous reduction in cell cost, initially by improving the economics of silicon wafer-based technology and then by switching to an inherently much lower cost 'thin-film' approach.

In the thin-film approach, a thin layer of the photoactive semiconductor is deposited onto a supporting substrate or superstrate, such as a piece of glass, as shown by the diagram in Figure 10. This approach is inherently much cheaper than using a self-supporting silicon wafer-based cell such as shown earlier in Figure 1 (page 8). Not only is a large amount of material saved, but the unit of manufacture, instead of being a 10 cm x 10 cm cell becomes a module that may be 100 times or more larger in area, lowering manufacturing costs.

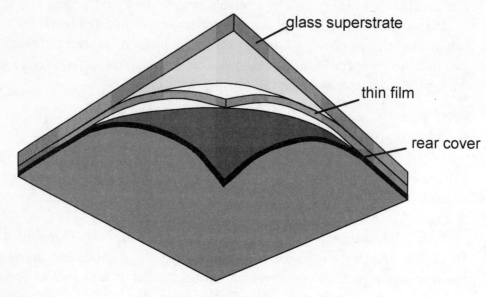

glass superstrate

thin film

rear cover

Figure 10
Thin-film layer of solar cell material deposited onto a supporting glass superstrate.

The plans of the solar cell industry were embraced by the US government, which set up a well-coordinated program to achieve these objectives. The program was given a further boost with the election of Jimmy Carter to the US presidency in 1977.

The program nurtured the birth of the infant terrestrial solar cell industry by a series of demonstration projects, targeting potential applications of photovoltaics. A particularly significant outcome from this program was the development of reliable low-cost techniques for packaging the silicon wafer cells into modules, and research into methods of lowering the cost of the silicon material used in the cells. By the 1980s, this program had helped create a small but stable terrestrial solar cell industry capable of supplying high quality product to the key market areas. These were mainly in remote regions where there was no reliable source of power, particularly for telecommunications, and in government subsidised demonstration programs.

Solar cells lose out

Jimmy Carter's administration, with its very supportive attitude to renewable energy was followed by Ronald Reagan's administration, which appeared antagonistic to the whole idea. Resources were shifted to the Strategic Defence Initiative, a grand plan to provide an umbrella defence against any incoming missile, colloquially known as 'Star Wars'. The funding for renewable energy decreased dramatically.

The consequences were felt almost immediately in the USA. University researchers were forced to switch to non-solar fields to get the funds needed for postgraduate student stipends and for their research programs. Many of the first class researchers that had been attracted to solar research during the Carter administration moved to other fields and have never returned. Other countries around the world, almost without exception, 'followed suit' and reduced the size and scope of their renewable energy programs. This complacency was shattered by a major nuclear reactor incident in Russia.

Solar cells make a come back

Nuclear reactors may provide a fine way of generating electricity as long as they are very competently designed, operated and maintained; the expertise needed to specify, build and operate them is not diverted to the development and proliferation of nuclear weapons; and suitable means can be found for storing nuclear waste over the approximately 10 000 years when high levels of radioactivity will remain.

The consequences of not meeting even some of these conditions were dramatically brought to the attention of the world and the European community, in particular, by the nuclear accident in Chernobyl in April, 1986. Apart from more dire consequences in the immediate vicinity, within days a cloud of radioactive material spread over northern Europe threatening crops and livestock. The incident stiffened resolve against the use of nuclear reactors in Europe and strengthened programs seeking to develop more acceptable means of electricity production.

We are starting to see the consequences. In 1998, 6 per cent of all electricity produced in Denmark was generated by wind energy, well on the way to the near-term target of 25 per cent. In Germany, wind energy accounted for 1 per cent from a base of almost zero only a few years earlier. As well as these initiatives, the Chernobyl nuclear accident revitalised research programs in photovoltaics as a longer-term way of meeting electricity requirements (the solar resource is about 200 times larger than wind, wave, tidal, hydroelectric, biomass and geothermal resources combined).

Since Chernobyl, there has been growing concern about the use of nuclear power, for example, in the reported use of civilian nuclear reactors in military nuclear weapon programs in Iraq and Pakistan, and changing economic and political circumstances such as in the former Soviet Union leading to increased risk from inadequate maintenance of nuclear plants. The smuggling of nuclear material, presumably for clandestine purposes, is also widely reported. Proliferation of nuclear reactors can only lead to increased problems of this type in the future.

With the photovoltaic programs put back on track by the Chernobyl nuclear scare, further urgency has been added by increasing scientific evidence for the severe environmental impact of the traditional ways of generating electricity.

Global warming: the greenhouse effect

The use of fossil fuels to supply the world's energy requirements has always been a messy business. Acid rain and petrochemical smog are two well-known consequences with enormous social and environmental costs. These costs are generally borne by the community as a whole rather than by the fossil fuel burners.

In the 1990s, scientific and public concern grew rapidly over a further conse-quence of the use of fossil fuels to supply the world's energy. It has been known for some time that carbon dioxide levels in the atmosphere are increasing as a direct result of human activity. There is also little doubt that the earth's surface temperature is steadily rising with an increase of 1°C documented over the last century. So far, 1998 is the hottest year on record with the previous hottest being 1997. It now seems almost certain that these facts are related and that the tem-perature rise seen over recent decades is a direct consequence of the release of car-bon dioxide and related gases into the atmosphere. BP Chief Executive Officer, John Browne sums up the current situation (Street, (1999), see Chapter 1):

> Of course, the science of climate change is ... provisional and perhaps always will be. But there is mounting evidence that the concentration of car-bon dioxide in the atmosphere is rising and the temperature of the earth's surface is increasing. There are large areas of uncertainty — about cause and effect and about consequences. But it would be unwise and potentially dan-gerous to ignore the mounting concern.

If ongoing increases in atmospheric carbon dioxide levels were left to continue unchecked, the level would double sometime around 2050 with estimates of likely temperature rise being an additional 2°C above present temperatures.

This could have quite dramatic effects upon the climate. The sea level would also rise due to melt-back of the ice caps in the polar regions, flooding many highly populated areas. The frequency of extreme climatic events would also be expected to increase.

The above concerns have triggered international action at the highest level. The Kyoto Agreement formulated in 1997 saw countries committing to reductions in greenhouse gas emissions by 2010 ranging from 18 per cent below 1990 levels in the case of Germany to small increases in other cases such as an 8 per cent increase in the case of Australia. To meet even the least challenging of these commitments, drastic changes in energy use are required. This has stimulated activities likely to reduce greenhouse gas emissions, including the accelerated development of photovoltaics, which holds such promise for a massive impact in the longer term.

In many countries, the public has also become involved in helping to reduce negative environmental impacts. A range of initiatives to promote solar cell use have been launched including 'rate-based incentives' and 'green-pricing' schemes.

Rate-based incentives and green-pricing

Starting from the tiny town of Burgdorf in Switzerland in 1991, schemes to promote the use of solar energy have spread around the world. These schemes allow members of the public concerned about environmental issues to help contribute to their solution. A range of schemes have been explored based on individuals or whole communities opting to pay more for electricity generated in an environmentally responsible way.

In the Burgdorf and Aachen rate-based incentives model, customers within a utility district vote on whether or not they wish to support the deployment of renewables by a small surcharge of about 1 per cent on their electricity bills. The extra revenue is used to pay a premium for electricity generated from privately owned renewable sources. Such schemes work best in cities where municipal utilities are responsible for power supply and local politicians have the power to implement the schemes. This is a common situation in Germany where over 8 per cent of the population now have access to such schemes. These schemes accounted for 30 per cent of all photovoltaic product sold in Germany in 1997.

In 'green-pricing' schemes, individual customers who want their electricity supplied from sustainable sources pay a premium. The electricity supplier then ensures that an amount of electricity matching the customer's quota is supplied to the electricity grid network from a renewable source.

A range of other schemes have also been implemented. Zurich introduced the 'solar stock exchange' where a separate market is established for solar generated electricity. The municipal utility calls for bids for privately generated photovoltaic electricity and selects the most attractive offer. A systematic discussion of these schemes can be found in the references for further reading.

Growth of the solar cell industry

By the early 1980s, driven by the oil crises of the previous decade, the solar cell industry had become reasonably mature. Processes for producing cells and modules, such as in Figures 1 and 2, were now largely standardised across the industry.

The industry has grown steadily since then, with growth since 1990 shown in Figure 11. Up until 1996, the industry grew at a healthy rate of 15–20 per cent per year. In 1997 and 1998, growth was a more explosive 20–40 per cent, driven by the rapidly increasing use of photovoltaics in private residences in urban areas. Over 10 000 systems like this were installed in 1998 alone, with this number expected to double in 1999. More than three million urban home systems are due to be installed by 2010, based on targets announced by governments in Europe, Japan, the United States and Australia.

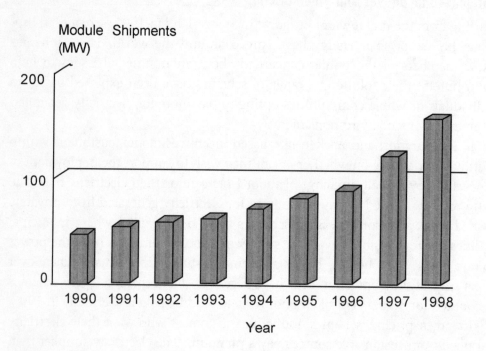

Figure 11
Growth in the shipment of photovoltaic modules since 1990 measured in megawatts (MW) of peak electricity generation capacity.
Source: PV Insider's report

The selling prices for solar cell modules have decreased steadily as production volumes have increased. Figure 4 (page 11) shows how the average wholesale price has reduced since the late 1970s.

Interestingly, the historical data points lie very close to being on a straight line when plotted in the way shown in Figure 4. The more of something that is produced, the cheaper it usually becomes. The slope of the straight line in Figure 4 quantifies how quickly this is happening in the case of solar cells. The straight line fit to the data shows that, for every doubling of the total number of solar modules produced by the industry, the cost drops to 77.5 per cent of the initial value. If this trend continues and the industry continues to grow at 30 per cent per year (as it has since 1996), photovoltaics will reduce to 50 per cent of their present price in 8 years. If the industry grows at only 20 per cent per year it will reach this price in 11 years.

If the residential programs mentioned meet their targeted goals, the accumulated production by the photovoltaic industry would be in excess of 10 gigawatts by 2010, compared to the value of 900 megawatts at the end of 1998. If the rate of price decrease documented in Figure 4 is maintained or increased by new technology, cell prices would decrease to between one-third to one-half of their 1998 costs over this period.

* * *

Since the first silicon cells were made quite by accident in the 1940s, the technology has progressed by fits and starts. During the 1950s and 1960s, it received a huge impetus from the need for a reliable and durable power source for use in space. The oil embargoes of the 1970s stimulated the interest in land-based use. Nuclear accidents at Three Mile Island and Chernobyl in the 1980s followed by increasing awareness of the impact of greenhouse gas emissions in the 1990s stiffened resolve to develop solar cell technology to its full potential.

Increasing production of cells has seen a corresponding reduction in their cost, with surprisingly close correlation between these two factors. Direct public involvement in the purchase and use is driving the present unprecedented growth in demand for these cells, which will make them much cheaper over the coming decade.

Further reading

Gabler, H., Heidler, K. and Hoffman, V.U. (1998), 'Grid connected photovoltaic installations in Germany — the success story of green pricing and rate-based incentives', *Second World Conference and Exhibition on Photovoltaic Solar Energy Conversion*, July, Vienna, Austria, pp. 3413–17.

Haas, R. (1998), 'Financial promotion strategies for residential PV systems — an international survey', *Second World Conference and Exhibition on Photovoltaic Solar Energy Conversion*, July, Vienna, Austria, pp. 333–38.
(Both papers give a good survey of the range of schemes promoting residential use.)

Perlin, J. (1999), *From Space to Earth: The Story of Solar Electricity*, AATEC Publications, Ann Arbor. (The first book to deal exclusively with the history of photovoltaics.)

Riordan, M. and Hoddeson, L. (1997), *Crystal Fire*, Norton, New York. (Gives an interesting and swiftly moving account of the birth of the semiconductor industry.)

4

TECHNOLOGICAL EVOLUTION: THE THICK AND THE THIN

Ideal solar cell features

For solar cells to be widely used, the three key technical requirements are low manufacturing cost, high reliability and high conversion efficiency of sunlight into electricity. Most of the research and development work in photovoltaics addresses one or more of these three requirements.

As discussed, solar cells have previously been too expensive for widespread use, although costs have decreased steadily as more cells are made (see Figure 4, page 11). The way the cells are made has also steadily improved, although there is room for further improvement.

Reliability has been one of the technology's great strengths. Solar cells could last forever since there is no wear-out process. With well-designed packaging, an operating life longer than that of traditional electricity generating equipment would be expected in the long term. Manufacturers now offer product warranties of up to 25 years, as earlier noted. It is most important that this reliability is not sacrificed in the effort to reduce costs.

The sunlight energy conversion efficiency fixes the area of solar module required for a given electrical output. Many of the costs of materials used in the system will also depend upon this area. While costs such as those of the glass sheet covering the cells are usually included in cell production costs, other costs such as those of mounting structures, inter-connection wire and so on have to be added. By estimating how low such added costs might eventually become (per unit area), energy conversion efficiency above 10 per cent and preferably above 15 per cent is thought to be necessary to keep these extra costs manageable in the long term.

Solar cell ingredients

Most of the solar cells sold up until now use the 'silicon wafer' technology mentioned in Chapter 1 in which the silicon wafer forms the bulk of the cell.

The starting material for these wafers could be sand, although purer quartz is normally used (both sand and quartz are a compound of silicon and oxygen). Heating with carbon extracts the oxygen, leaving silicon. This silicon is purified (by 'fractional distillation' as in oil refining, by forming a volatile silicon compound and re-extracting silicon after this compound is purified). The purified silicon is then melted and solidified onto a pre-formed 'seed' crystal that acts as a template for the growth of a perfect crystal as it is slowly drawn from the molten silicon. A cylindrical ingot of silicon is the end result, as shown in Figure 12. Once formed, the ingot is sliced into individual wafers. Sometimes, it is partly 'squared-off' by sawing off the edges and recycling these.

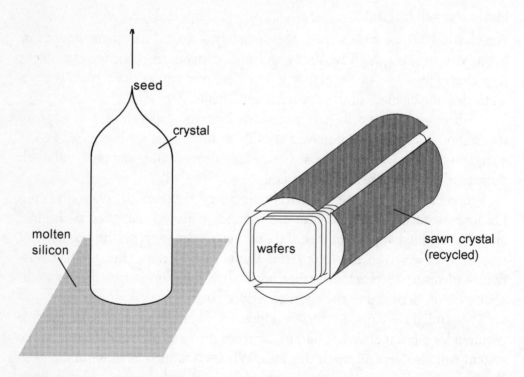

Figure 12
Growth of a cylindrical crystal of silicon from a melt onto a small silicon seed and the slicing of thin wafers from this crystal (squaring-off is optional).

The next step is to attack the wafer's surface with chemicals to remove a damaged layer formed during the wafering step. A second chemical attack produces tiny pyramids over the wafer surface that are far too small to be seen by eye and are shown greatly exaggerated in size in Figure 13. These miniature pyramids reduce reflection from the silicon surface by causing light to be reflected downwards, when it first hits a pyramid side.

Next comes the p-n junction. The wafer starts 'p-type', since a small amount of boron is added deliberately to the melt. The 'n-type' side of the p-n junction is formed by allowing phosphorus to seep into the solid wafer surface at high temperature, overriding the boron in this surface region.

The metal contacts are then applied. In the standard approach, this is done by a 'screen-printing' step similar to that used for making customised T-shirts. A paste containing metal particles is squeezed through a patterned screen onto the surface of the cell (the screen controls both the thickness and shape of the resulting pattern). The cell is then heated to drive off almost everything from the paste but the metal. The best results are obtained when the metal is silver.

A cross-sectional view of the final cell is shown in Figure 13. Standard cells convert between 11–15 per cent of the incident solar energy into electricity. When packaged into modules, the energy conversion efficiency of the module is a little lower due to space wasted by frames and the gaps between the cells, with module efficiencies lying in the 10–13 per cent range.

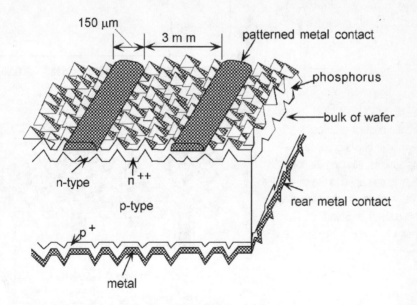

Figure 13
Traditional 'screen-printed' silicon wafer-based solar cell with features shown greatly magnified (and not to scale). This is a close-up view of the cell shown in Figure 1 (page 8).

How to improve efficiency

Leaving aside the need for low cost, Figure 14 shows how the efficiency of the 'best ever' laboratory silicon solar cell has risen since the first-ever devices were made in the 1940s to about 25 per cent — nearly double that of the standard commercial product. What scope is there for getting these recent improvements into commercial product? The answer is — plenty.

Buried contact cell

A new commercial solar cell structure known as the 'buried contact' solar cell, has many of the high efficiency features of the best 'lab' cells (see Figure 15).

Figure 14
Evolution of the energy conversion efficiency of the best laboratory silicon solar cells since 1940 (the author's group at the University of New South Wales has been responsible for much of the recent improvement , as shown by the black data points).

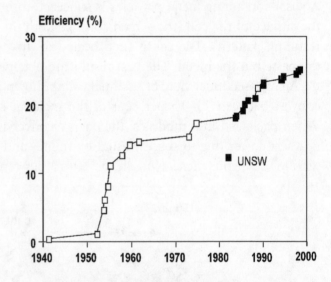

Figure 15
The buried contact solar cell where a laser blasts grooves in the top surface, allowing the top contacts to be buried in the silicon (cell invented by the author and his colleague, Stuart Wenham).

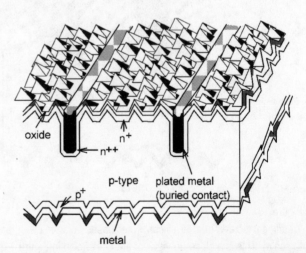

Instead of lying flat across the surface and blocking sunlight, the top contacts are buried into deep grooves within the cell surface. Using the same quality silicon wafers, the cells give 20–30 per cent extra power at no extra cost. Buried contact cells are now available through BP Solarex as the 'Saturn' product line. Module efficiencies are in the 14–15 per cent range (with cell efficiencies of 16–18 per cent), the most efficient available commercially. By boosting efficiency without increasing processing cost, the overall economics are correspondingly improved.

Ribbon substrates

Clearly, producing a cylindrical ingot of silicon and then slicing it into wafers is not a particularly refined or efficient way of producing the large areas of silicon needed to generate substantial power. Forming silicon directly as large area sheets or ribbons would be more sensible.

Commercially, the most advanced ribbon process is known as the 'edge-defined film-fed growth' (EFG) method shown in Figure 16. In this approach, molten silicon moves up between the faces of a carbon die, forming a ribbon as it is drawn from the top of this die. Improved variations that produce a nonagonal tube of silicon are now used in commercial production. Cells are made in much the same way as on standard wafers, using wafers cut by laser from this ribbon silicon.

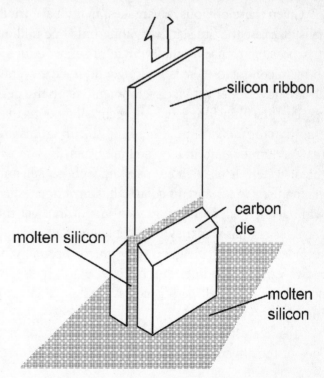

Figure 16
The basic process used to form ribbon silicon by the EFG process (the process has now been refined to produce a nonagonal tube of silicon).

Thin films

The above approaches will help to reduce cell prices. However, the greatest potential for cost reduction is to get away entirely from the idea of self-supporting solar cells and instead deposit cells straight onto a stronger supporting layer.

Figure 10 (page 24) illustrates how a thin layer of the photoactive material is deposited onto a sheet of glass. This means that the amount of semiconductor can be dramatically reduced. Another advantage is that the size of the item manufactured is no longer limited to that of a silicon wafer. Instead, it is determined by the size of the glass sheet, which may be over a hundred times larger in area. Both these features make thin-films the 'ultimate solution' for photovoltaics.

In large production volumes, the costs of fabricating thin-film cells will approach those of the materials used. For the structure shown in Figure 10, the major material cost is that of the glass sheet. Ultimately, therefore, photovoltaics can be very low in cost, perhaps only a few times the cost of such a sheet of glass, making it possible to produce electricity at very low cost. (Many studies predict that thin-film solar cells could eventually be made for less than US$30 per square metre. At 15 per cent efficiency and with 1 kilowatt per square metre of sunlight available, this corresponds to less than $200 per kilowatt of electricity capacity resulting in electricity generation costs less than those using conventional fuels.)

Given these obvious advantages, many have tried to develop thin-film solar cells. Because the films are very thin, only one millionth of a metre in thickness, it is possible to use a whole range of semiconductors apart from silicon since the semiconductor cost is no longer an issue. Several different thin-film materials are now under active development, as briefly described below.

The first thin-film solar cell technology onto the market was based on silicon in amorphous form, where the silicon atoms are arranged more randomly (as in a liquid) than in a crystalline wafer. If you own a solar-powered calculator or watch, it most likely uses amorphous silicon thin-film cells. Unlike the silicon used in a standard solar cell, amorphous silicon is prepared very simply, without melting, so the silicon atoms do not get the chance to arrange themselves as regularly as they would like. Not all amorphous silicon atoms are surrounded by four neighbours, which normally would give poor results. However, by flooding the material with hydrogen during its preparation, acceptable solar cell performance is obtained (the hydrogen replaces the missing neighbours).

It is interesting to see how amorphous silicon modules are made, to give an idea of the advantages of the thin-film approach. Starting with a glass sheet, a thin layer of a 'transparent conductor' is deposited onto this sheet (the transparent conductor is a good conductor of electricity but lets light through). Using a laser, this layer is then patterned into strips that define the final location of the cell. The photoactive amorphous silicon semiconductor layers are then deposited and also patterned into strips using a laser, but these strips are slightly offset from the transparent conductor layer.

Finally, a metal layer is deposited with similar patterning, again offset as shown in Figure 17. By this series of offsets, individual cell areas defined by the laser patterning are automatically interconnected. The sheet of glass ends up covered by a number of solar cells automatically connected together in series. This completely eliminates the steps involved with connecting individual cells together in a traditional silicon wafer-based module (Figure 2, page 8).

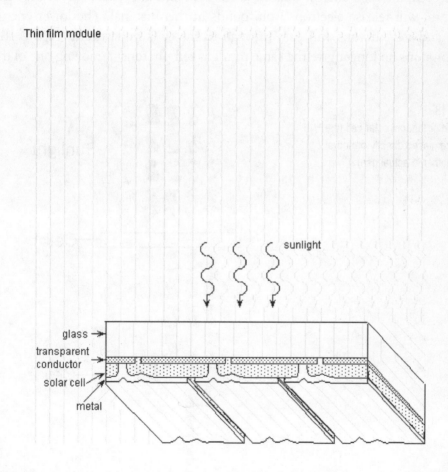

Figure 17
Method used to fabricate a thin-film amorphous silicon module. The three cells shown are automatically connected together.

Although this approach has obvious attractions due to its potential for low cost, the big challenge has been to get high efficiency, due to the poor quality of amorphous silicon material even with the boost provided by the hydrogen. Another complication is that this boost tends to undo itself during exposure to light. To make a stable cell, the amorphous silicon has to be a lot thinner than you really want, if the whole cell is to remain active as the material degrades.

One way around this problem is to stack two or more solar cells on top of one another. In this way, each of the cells can be kept thin and relatively stable, while the total thickness is large enough to capture most of the light.

It turns out that you can do even better if the cells are made from different semiconductor materials, each needing different photon energies to release electrons from bonds. If the material requiring the highest energy photons is used for the top cell, the high energy photons at the blue and ultraviolet end of the solar spectrum will release electrons from bonds in this material. The lower energy green and red photons will pass through to the cells underneath (Figure 18). Blue photons no longer give the same result as red photons, removing one of the

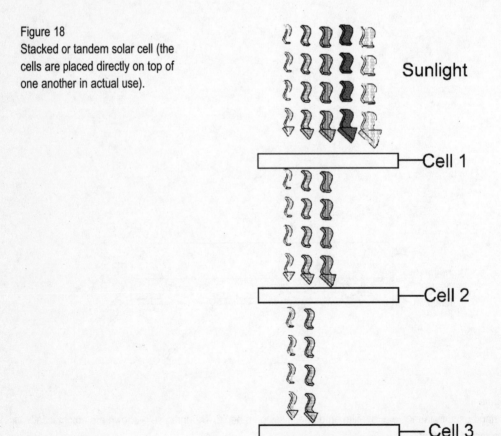

Figure 18
Stacked or tandem solar cell (the cells are placed directly on top of one another in actual use).

Sunlight

Cell 1

Cell 2

Cell 3

major loss mechanisms mentioned in Chapter 2. In the stacked cell approach, the blue photons are used in material where they can be converted efficiently. So are the red photons, reducing losses compared to a single cell device.

Using different semiconductors

The energy required to release an electron from a bond in amorphous silicon can be controlled relatively simply by mixing the silicon with a very closely related semiconductor, germanium. This is exactly what is done in the best amorphous silicon cells. An amorphous silicon cell is uppermost and converts the blue wavelengths. Underneath lies a cell made from a mixture of amorphous silicon and germanium, which converts the green wavelengths. A third cell underneath the top two, with even more germanium added, converts the red wavelengths.

Even with this sophistication, the resulting performance is quite modest since the quality of the amorphous material is so marginal. The best commercial modules of this type have efficiency in the 6–7 per cent range, less than half that of the best wafer-based silicon product. The overall economics are still promising, however, due to the material savings and the manufacturing advantages mentioned. Another strong point is that almost the whole module is coated with cell material that gives a more attractive, uniform appearance than with the standard module (Figure 2, page 8).

The use of germanium in these cells also brings another issue to the fore — that of the available resources of material used in the cells. Unlike silicon, germanium is very rare with known worldwide reserves of a few thousand tonnes. If all these reserves were used for cell production, it would still not be possible to make enough cells to contribute significantly to the world's electricity requirements, not to mention the total energy requirements. These considerations aside, the available germanium would be sufficient to tide the industry over the next decade or two. By then, an alternative approach, such as replacing the germanium with another closely related element, tin, may alleviate this problem in the longer term.

Other materials

Many other materials, involving compounds formed by combining two or more elements, are also suitable for improving the efficiency of thin-film cells, however they have drawbacks that may limit their use.

One such compound is that formed by cadmium and tellurium, cadmium telluride (CdTe). Cells can be made very simply from this material. A major disadvantage is that the material is toxic, which could limit its potential in a market driven primarily by environmental concerns.

Another material that has given good results is based on the ternary compound, copper indium diselenide ($CuInSe_2$), often referred to simply as 'CIS'. Additional elements are also often mixed into the CIS, particularly gallium (Ga) and sulphur (S), making it an elemental *potpourri* — CIGSS. Indium is as scarce as germanium, which may create a similar resource problem in the longer term.

Another type of solar cell mimics photosynthesis. Light is absorbed in a photosensitive dye attached to the surface of semiconductor material (TiO_2). In this case, the mobile electron is released from the dye rather than from within the semiconductor. The electrical circuit is completed by a conductive liquid electrolyte. This liquid turns out to be a disadvantage, since it is difficult to seal liquids within the module for the long operating life required. The advantage of the dye, however, is that it absorbs only a narrow range of photon energies. These energies could lie in the infrared range of photon energy, meaning that these cells could be transparent to visible light while still producing electricity. The advantage is that they could be used in completely clear windows that could also generate power.

Despite all these innovative options, it appears that the next generation of thin-film photovoltaic solar cells could be based on something much closer to the standard silicon-wafer cell. Recent research shows that thin layers of silicon can be deposited directly onto glass with properties quite similar to the wafer material.

This might seem quite an obvious thing to do, if you want to get away from the use of wafers to reduce material costs. One problem has been that normal silicon is a weak absorber of sunlight compared to the other semiconductors mentioned above (including even amorphous silicon). Much thicker layers were therefore thought necessary to make a decent cell, complicating the situation enormously since it is more difficult and expensive to deposit thick layers of good quality silicon than thin layers.

By arranging to trap light into the cell by disorienting it once it is inside, these thin layers can appear as if they are up to 50 times thicker than their actual thickness. Combining this new 'light trapping' feature with solar cell designs that take advantage of the lower quality material in such films, modules not significantly lower in performance than their silicon wafer counterpart seem likely to be developed (Figure 19). These silicon thin-film cells offer the low cost of the thin-film approach with the established performance and reliability of the silicon wafer approach.

***

The solar cell of the future should become more cost effective through increased manufacturing volume combined with technological improvements such as buried contact, ribbon and thin-film cells. This means that the historical trend in price reduction shown in Figure 4 (page 11) should be able to be maintained for the coming decade or two. We can look forward to a future of steadily decreasing cell prices and an expanding range of uses.

Figure 19
A silicon thin-film solar cell module produced in pilot production by the Sydney company, Pacific Solar Pty Ltd. Features include depositing of the silicon film directly onto a glass sheet, a light-trapping scheme to make the film appear much thicker than its physical thickness and a new cell design suited to the lower cost silicon used in the cell.

Further reading

Kazmerski, L.L. (1998), 'Photovoltaics: a review of cell and module technologies', *Renewable and Sustainable Energy Reviews*, Pergammon, vol. 1, pp. 71–170. (Contains a good scientific review of the different cell and module technologies plus an extensive bibliography.)

Partain, L. (ed.) (1995), *Solar Cells and Their Applications*, Wiley, New York. (Gives an up-to-date account of the different solar cell technologies written by experts.)

5

HOW CAN SOLAR CELLS BE USED?

Evolution of solar cell use

As the cost of solar cells decreases, the range of possible uses expands. Twenty-five years ago, solar cells were very expensive and were used only for generating small amounts of electricity in areas remote from other sources of electricity. An extreme example is the use on spacecraft.

Closer to home, cells have been widely used to power telecommunication equipment in remote areas. An early system, for example, was the telecommunications link between Tenant Creek and Alice Springs in the Australian outback. When dialling Alice Springs, the telephone signal passes through 13 solar-powered stations that relay the signal across the hundreds of kilometres between these towns. Telecommunications has formed the backbone of the terrestrial solar cell industry until quite recently.

With decreasing costs, the range of possible uses has increased — a trend that will continue. Twenty-five years into the future, the cells are expected to be only a fraction of their present cost and be able to compete in the most demanding applications, such as side by side with large conventional electricity generators. In between these past and future extremes, there are many other applications, some of which are suggested in Figure 20.

Current uses

In remote area power supplies, photovoltaics are usually used with 'lead-acid' batteries, similar to the normal car battery, to store electricity overnight and for periods of low sunshine. In remote areas, photovoltaics can also form part of a larger 'hybrid' system (explained later in this chapter) complemented by other power sources such as diesel generators, as well as battery storage. Remote area power supplies are used both for 'professional' telecommunication systems and for providing electricity for isolated homes. Systems of this type can provide power to people in remote locations in the developing world, as discussed further in Chapter 8.

In industrialised countries, the largest applications are those where the photovoltaic system is connected to the existing electricity distribution wires. Large centralised photovoltaic generating plants have already been mentioned as a future prospect (Figure 3). This is the most demanding application since the only value credited to the photovoltaic plant is the electricity produced. However, there are other grid-connected applications (described below) where the photovoltaics give extra benefits.

One is quite technical, but involves using photovoltaics to postpone spending to upgrade parts of the grid network that are becoming overloaded (as indicated on the right-hand side of Figure 20). Here, advantage can be taken of the 'modularity' of photovoltaics, that is, the ease of installing small amounts anywhere within the grid network. By installing photovoltaics near the area of increased demand, the need to upgrade the grid system supplying electricity to this area can be deferred. For example, there is then no need for additional transmission lines to get the power to this region or the need for upgrades to other equipment, such as transformers. In this case, not only do you get value from the electricity produced by photovoltaics, but also from the savings by postponing such upgrades.

The most important grid-connected application at the moment is the use of solar cells on the roofs of private homes. Here, the important point is that the price the home-owner must pay for electricity is several times higher than the cost to the power company operating the grid network to produce it (or purchase it) — basically, the difference between wholesale and retail electricity prices. The use on private residences is the most rapidly growing area of photovoltaics at the moment and Chapter 6 will explain why this is so. The closely related architectural use of photovoltaics is discussed in Chapter 7. Some of the other key applications are described below.

Remote power supplies

The simplest possible system for remote power is shown in Figure 21. All that is needed is a photovoltaic panel connected across a battery with a fuse to protect the battery from short-circuits. Any electrical load that can be driven by the battery can then be connected across it, without the battery going flat, if everything is properly sized.

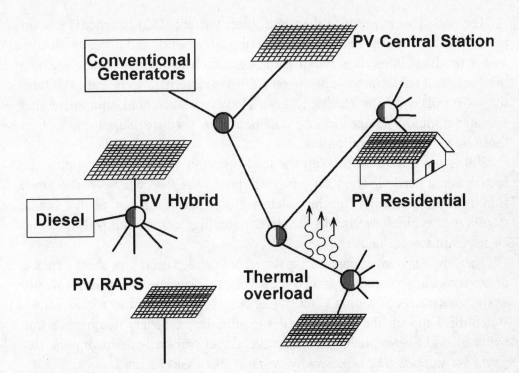

Figure 20
Solar cell applications. The photovoltaics (PV) remote area power supply (RAPS) systems represent
one extreme on this chart in terms of allowable PV costs, while the PV central power station represents
the other.

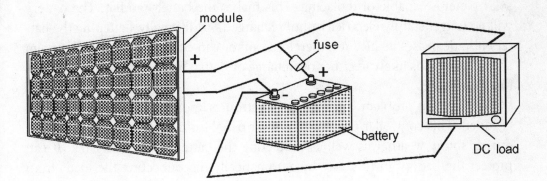

Figure 21
Simplest photovoltaic system for remote electricity supply.

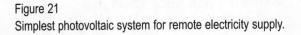

The solar panels produce the same direct current (DC) electricity as a car battery or the 'dry-cell' batteries used in transistor radios and other small consumer products. Most household equipment such as television sets, washing machines and refrigerators, operate on a different alternating current (AC) form that is supplied by the electricity grid. However, household equipment that operates from DC sources can be obtained from specialist suppliers, such as those selling camping equipment.

If it is not important to keep the load operating without interruption, the battery is not needed. The solar panel will power the load whenever the sun is shining strongly enough and the system will stop when it is not. Such a simple system is possible, for example, for water pumping because pumped water can be stored in storage tanks.

The simple system in Figure 21 is not recommended except for short-term use or when carefully monitored as there are several problems with it. In sunny weather, the solar panel can supply so much electricity to the battery that it overcharges. When this happens, the acid and water mixture in the battery decomposes into hydrogen and oxygen, reducing the acid level and eventually destroying the battery if not stopped. The explosive hydrogen is also a safety hazard.

Some module manufacturers make what are called 'self-regulating' modules that have fewer solar cells than the normal 36 in them (32 or 33, instead) and, therefore, there is less chance of overcharging the battery (for the more technically minded, a few diodes in series with a standard module to reduce the output voltage will do the same job).

On the other hand, if there is not much sun or you have been too ambitious in your demand for electricity, more will be drawn from the battery than the solar panel is capable of replacing. This makes the battery go flat. The battery will no longer supply electricity until recharged. Additionally, flattening the battery also decreases its life. You therefore cannot win either way. A simple system as in Figure 21 is likely to either overcharge or flatten the battery — both reduce the battery life.

To solve this problem, a package of electronics known as a 'charge controller' is added. This controller prevents the solar panel from overcharging the battery during sunny weather as well as protecting the battery from going flat. It can protect the battery in this way by automatically disconnecting the loads from the battery — not always appreciated by the system users. Alternatively, an alarm light can be used when the charge in the battery becomes low and the users can take heed or not (like the engine overheating alarm in a car — you can keep driving if you don't mind running the risk of your engine 'seizing up').

Since electrical equipment is more easily obtained if it operates from AC electricity rather than DC, a further refinement is to add another package of electronics, known as an inverter, between the battery and the electrical load. The inverter converts the DC output of the battery to AC form.

Systems as shown in Figure 21 but including a charge controller can work very well and have been used in a wide range of applications. At one extreme are the systems used in telecommunications. These are very carefully engineered and operate at extremely high levels of reliability with little maintenance. In these systems, the size of the electrical load usually can be predicted accurately. Enough battery storage generally is provided to tide the system over about 10 days without sunshine. This is to cover periods of bad weather or some problem in getting power from the panel to the load, allowing time for a maintenance team to arrive. Even though solar panels have been expensive in the past, the cost of batteries for 10 days' storage is roughly comparable to panel cost. These systems are now so well established and their reliability so well proved that they are the automatic choice for many applications.

At the other extreme might be a system (as in Figure 21) used to power an isolated home where there is no conventional electricity supply. The home might be a holiday home or the family residence in remote areas, such as in outback Australia or in rural areas of the developing world. Here the system user ideally becomes familiar with the capability of the system and rations loads to ensure the batteries seldom go flat.

When substantial amounts of power are required, the amount of battery storage required for reliable operation can be daunting. Here, 'hybrid' systems come into their own.

Hybrid systems

In a hybrid system, the photovoltaic modules are combined with more traditional electricity generators to form a system with better overall features. An example is shown in Figure 22. As well as the photovoltaics and battery, a diesel- or petrol-fuelled electricity generator set is added to the system. This generator set removes the need for large amounts of battery storage, since it can be started up to take over the load or recharge the batteries, when needed. Various configurations are possible, ranging from systems where the generator set and photovoltaics are used almost completely independently, each supplying their own loads, to very highly integrated systems.

Depending on the relative size of the diesel- or petrol-fuelled set and the photovoltaics, there are different ways of thinking about the role of the different components in the system. For example, if the system is primarily photovoltaic, the generator set can be thought of as a way of reducing the amount and

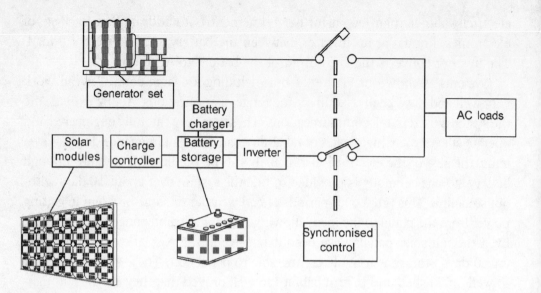

Figure 22
A hybrid system combining a diesel- or petrol-fuelled motor-generator set with photovoltaics to reduce the size of the battery storage. In the parallel system shown, the AC outputs of the generator set and solar system can be combined by the synchroniser. Either the solar cells or the generator set can charge the batteries or supply the load.

hence the cost of battery storage, since diesel- or petrol-fuelled generators are relatively cheap but have high running costs. These running costs are kept to a minimum if the generator set only needs to be used infrequently. A small petrol-fuelled generator is less expensive but less durable than a diesel generator and might be a good option in such cases.

On the other hand, the photovoltaics might be integrated into an already existing diesel system. In this case, the photovoltaics could be thought of as a way of saving fuel that would otherwise be consumed by the diesel generator. Again, the cost of the photovoltaic system can be quickly recovered in cases where it is difficult to transport diesel fuel and dispose of fuel drums, such as for island communities.

Grid-connection

The really large applications for photovoltaics in the developed world are those where they are connected to the electricity supply grid. This is challenging, economically, since the solar-generated electricity has to compete with that already available to the grid. Applications where photovoltaics add value, in one form or another, additional to the electricity they produce are therefore the most interesting for now.

No battery storage is required for grid-connected photovoltaics, at least while they supply less than 10–15 per cent of the total electricity used by the grid. This may seem surprising since, during bad weather, the photovoltaics would not generate much electricity and some 'back-up' would seem to be needed. In fact, the grid already must be operated in a way that adjusts to fluctuations in demand for electricity. These fluctuations in demand are similar in nature to fluctuations in supply produced by a solar system. No change in the way the grid already operates is required until the solar penetration becomes larger than the 10–15 per cent figure previously mentioned. The time when this will happen is probably one or two decades away in most countries.

The way the grid meets fluctuations in demand is by having what are known as 'spinning reserves'. Surplus conventional generators are kept 'idling' so that, if there is a sudden increase in the electricity demand, these can quickly supply the extra needed. These spinning reserves also protect against the unplanned loss of power from a conventional power generator such as caused by a fault at a large power station. Typically, the spare capacity available in spinning reserves would be about 30 per cent of the total load at any point in time.

Since storage is not required for photovoltaic use in a grid, the solar system can be very simple — all that is required is the solar panel and an inverter, previously mentioned. This inverter converts the DC output of the panel to AC and also provides protective features. For safety reasons, whenever the conventional power to the grid is turned off (for example, for maintenance or repair of the grid lines), the inverter has to shut down so that no power is sent from the photovoltaics.

Photovoltaics on the roofs of homes or integrated into the facades of buildings are the most common grid-connected uses to date. These applications take full advantage of the modularity of photovoltaics — systems can be small and still work effectively. The number of systems like this will grow from the present tens of thousands to millions over the coming decade. These residential and building integrated systems are discussed in more detail in Chapters 6 and 7.

An example of a different type of grid-connected system is one installed at Kerman in California in 1992. Growing demands for electricity in this region were putting increasing pressure on the electricity distribution system, in particular, a large transformer at the Kerman substation. By installing a 0.5 megawatt photovoltaic system on the load side of this transformer, the need for replacing it by a bigger one was postponed.

The Kerman project was experimental since its aim was to get firm data on the benefits of such an installation. A detailed study found other benefits apart from extending the life of the transformer, such as an increase in the amount of power that could be carried by the nearby power lines.

Several other large solar systems have been connected to a grid. Although most have been as demonstrations, the largest in the Southern Hemisphere, that at Singleton near Sydney, was installed to meet the 'greenpower' commitments of its owner, the utility Energy Australia. Twenty-five per cent of the extra revenue generated from this company's green-pricing scheme is invested in new photovoltaic systems (the other 75 per cent is distributed between wind, biomass and small hydro systems).

Powering the world

What happens once photovoltaics supplies more than 10–15 per cent of the electricity in the grid network? Can it go on from here and eventually supply most of the world's energy?

The amount of solar energy falling on the earth is so large that there is no argument against such widespread use based on available sunlight and land resources. As previously mentioned, only 1 per cent of the world's barren area would need to be covered by 15 per cent efficient photovoltaics to supply all the world's forecast energy consumption in the year 2050. Over 1 per cent of the surface area of the United States is already covered by roads and by buildings. Covering 1 per cent of the earth's deserts with photovoltaics would be a big job, but within engineering capabilities. We also do not need to do it overnight, but can pace ourselves if we start early enough.

Once more than 10–15 per cent of the world's electricity is supplied by photovoltaics, some changes are needed. Some form of energy storage is required. On a much larger scale, this storage would replace the battery in the hybrid RAPS (remote area photovoltaics system) mentioned earlier.

An enormous amount of effort over the last 30 years has gone into developing improved batteries for electric vehicles and also for large-scale electricity storage. Over the next two decades, a more cost-effective battery may emerge than present lead-acid batteries. Batteries are therefore not completely out of the question.

However, there are two methods currently used for large scale electricity storage. One is 'pumped hydro' storage. This involves using surplus electricity to pump water uphill. The water then runs downhill driving hydro-electricity generators to supplement supply when required. Many large-scale pumped hydro storage systems like this are operating worldwide. The main disadvantage is the relatively low 'round-cycle' efficiency. At best, only about 60 per cent of the electricity is extracted compared to that used in pumping the water uphill. The environmental and social consequences of any large-scale hydro-electric project also need to be considered carefully.

The other large-scale but less common current method of storage is to use excess electricity to compress air in huge underground caverns. The air can be allowed to expand to drive turbines that again generate electricity when needed. The use of large spinning fly-wheels also may be feasible.

Global networks

Another imaginative approach largely removes the need for storage. This is the GENESIS scheme suggested by Dr Yukinori Kuwano, the inventor of the solar-powered calculator. GENESIS stands for 'global energy network equipped with solar cells and international superconductor grids'. The scheme is made credible by the observation that electrical distribution networks are becoming more and more interconnected. For example, Canada is connected to the United States; Europe is connected from one country to the other.

This increased connectivity stems from the reducing costs of transmitting electricity over long distances. If this trend continues or is accelerated by new technology such as superconducting transmission lines, it is not difficult to imagine these international connections becoming intercontinental. With a globally interconnected electricity network, the need for storage is greatly reduced. At any point in time, somewhere in the world, the sun will be shining onto solar cells. The electricity they generate can be pumped via the global network to regions where the sun is not shining.

An even more futuristic idea involves large solar cell systems in space that beam the electricity generated back to earth as microwaves free from the vagaries of weather and from the constraints of day and night. It is not clear whether these advantages would ever offset the much higher costs of deployment and maintenance.

Chemical storage

However, an entirely different route seems more practical and would also allow photovoltaics to supply energy in forms other than electricity. This involves storing the energy captured by solar cells in chemical fuels. For example, electricity can be used to decompose water (H_2O) into hydrogen and oxygen (such as when overcharging a lead-acid battery). Hydrogen can be used as a fuel in its own right to power vehicles such as aircraft and motor vehicles. It could also serve as an industrial fuel in the way that 'natural gas' (a mixture of compounds of hydrogen and carbon) is presently used. Hydrogen can also be converted back into electricity using 'fuel cells', where such fuels react directly to give electricity rather than being burnt. Although technology for this hydrogen cycle is already available, the problems at the moment are high costs and again, relatively low round-cycle conversion efficiency. Improvements might be expected in both areas over the next two decades.

Hydrogen is not the only option for a chemical storage fuel. A Japanese group, Research Institute of Innovative Technology for the Earth (RITE), is exploring the use of photovoltaic-generated electricity to convert carbon dioxide to methanol in sunny areas of the world such as in Australia (Figure 23). The fuel could then be shipped to the point of use, in this case, Japan. When the methanol is used to provide energy, carbon dioxide is produced — this would then be shipped back to the photovoltaic site, re-charged and sent back on its cycle. Incorporated into such a chemical cycle, there would be no barrier to photovoltaic energy supplying all the world's energy requirements, both electrical and non-electrical, provided costs could be made sufficiently low.

***

Photovoltaics has come a long way over the last two decades in terms of its suitability for widespread applications. We will see more in use over the next two decades. It seems to be entirely feasible that photovoltaics could supply almost all the world's energy requirements, in the longer term, say, in the 2050–2100 timeframe.

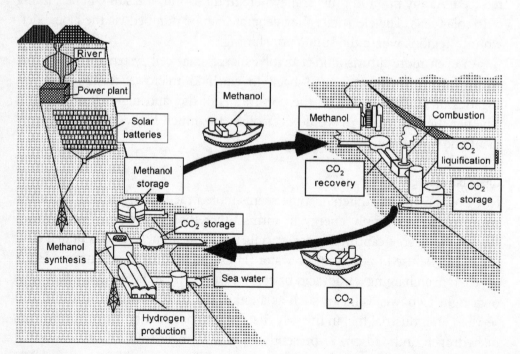

Figure 23

A photovoltaic powered, methanol-based chemical fuel scenario under investigation by the Japanese group, RITE (Research Institute of Innovative Technology for the Earth).

Further reading

Research Institute of Innovative Technology for the Earth (RITE), undated brochure, *Project for Chemical CO$_2$ Fixation and Utilization*. (Available from RITE, 9–2 Kizugawa-dai, Kizu-cho, Souraku-gun, Kyoto 619-02 Japan.)

Zweibel, K. and Green, M.A. (eds) (2000), 'The future of photovoltaics', special millennium issue of the journal, *Progress in Photovoltaics*, January, Wiley. (Contains a collection of papers outlining views on the future of photo-voltaic technology and applications by experts in the field.)

6

PHOTOVOLTAICS ON THE FAMILY HOME

Sell by day, buy by night

In the past, urban home owners have not always had much choice in the way electricity is supplied to their homes. All this is changing. Photovoltaics makes it possible to generate your own electricity on your rooftop, although presently at a cost premium. A rapidly increasing number of households worldwide are choosing this option with several million homes likely to be photovoltaic-powered by the year 2010.

The home remains connected to the power lines but no storage is required on-site, only a box of electronics (the inverter) to interface between the photovoltaics and the grid network. Figure 24 illustrates the system. During the day, when the home may not be using much electricity, excess power from the solar array is fed back to the grid, to factories and offices that need daytime power. At night, power flows the opposite way. The grid network effectively provides storage. If the demand for electricity is well matched to when the sun shines, as in many parts of the world, electricity produced during the day is more valuable than that consumed at night. This occurs in places like California and Japan where there are large air-conditioning loads and heating loads are small or provided by other fuels.

Early days

The first systematic exploration of the use of photovoltaics for use on homes began in the USA during the Carter administration. A well conceived program started by the siting of a number of 'residential experiment stations' at selected locations around the United States, representing different climatic zones. These

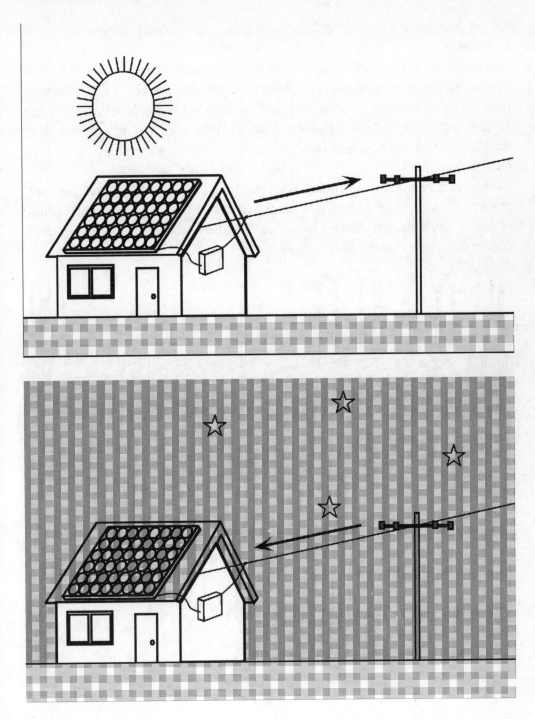

Figure 24
Residential use of photovoltaics — by day excess power is sent to the grid, and by night power is supplied to the home. (Drawing courtesy of Pacific Solar Pty Ltd)

stations contained a number of 'dummy' houses, each with a different solar system design.

Homes within the community close to these stations were monitored to see how well their energy use matched the energy generated by the station's dummy roofs. A small number of homes were also equipped with photovoltaics, prior to the planned installation of clusters of about 100 homes to confront, 'head on', the institutional and engineering issues arising from the widespread use of photovoltaics in this way.

The change in US government priorities in the early 1980s caused this program to flounder. One legacy is a small number of houses such as the Carlisle House in Massachusetts shown in Figure 25, which was probably the first grid-connected solar-powered house, worldwide.

Figure 25
The Carlisle House in Massachusetts which uses a roof integrated photovoltaic system to supply the electricity requirements of the home.

With the US effort dropping away, the Japanese Sunshine Project came to the fore. A large residential test station was installed on Rokko Island, a large artificial island near Kobe, beginning in 1986 (Figure 26). This installation consists of 180 'dummy' homes each equipped with its own 2–5 kilowatt photovoltaic system (about 20–50 square metres for each system). Some of the simulated homes (the huts in Figure 26), have their own electrical appliances inside, such as TV sets, refrigerators and air conditioning units, which switch on and off under computer control — providing a lavish lifestyle for their phantom occupants. For the other systems, electronics simulate these household loads. Also shown in Figure 26 are other pieces of equipment that can be configured to simulate different operating conditions within the electricity distribution network.

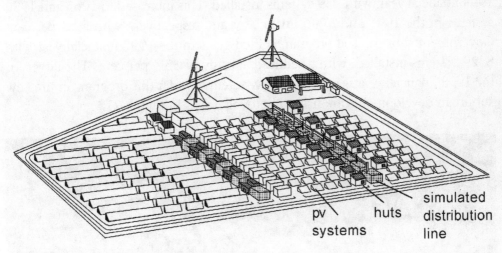

pv systems   huts   simulated distribution line

Figure 26
Rokko Island test station for residential photovoltaic systems. The simulated distribution line is used to test different situations within the grid network.

This test station has allowed the technical issues involved in using photovoltaics within the electricity network to be explored in a systematic way, under well-controlled test conditions. With no insurmountable problems identified, the Japanese have used the experience gained from this station to begin their own massive residential photovoltaics campaign, described below.

Meanwhile Germany began a very important '1000 roof program' in 1990 aimed at installing photovoltaics on the roofs of 1000 private homes. Large federal and regional government subsidies were involved, accounting in most cases for 70 per cent of the total system costs. The program proved immensely popular, forcing its extension to over 2000 homes scattered across Germany.

The success of this program had a number of effects. It stimulated other countries to launch similar programs as we will see below. It also helped the spread of 'rate-based incentive' and 'green-pricing' schemes (see Chapter 3) within Europe, particularly Switzerland and Germany, promoting the residential use of photovoltaics.

One million roofs

Japan's 'one million roof' program was prompted by the experience gained in the Rokko Island test site and the success of the German 1000 roof program. The initially quoted aims of the Japanese New Energy Development Organisation were to have 70 000 homes equipped with photovoltaics by the year 2000, on the way to 1 million by 2010. The program made quite a modest start in the 1994 financial year, with 539 systems installed. This increased to 1065 and 1986 systems in the 1995 and 1996 financial years, respectively, with a 50 per cent Government subsidy. The program shifted gears in the 1997 financial year with 8329 systems installed, with a reduced subsidy of $33\frac{1}{3}$ per cent. The target for the 1998 financial year was 15 000 new systems. Under this program, entire new suburban developments are using photovoltaics as in Figure 27.

Figure 27
Residential development in Japan using photovoltaics. (Photo courtesy of Resources Total System Co. Ltd)

Although the appearance of the installations on Rokko Island and those shown in Figure 27 is austere, some earlier Japanese initiatives show that aesthetics may well become a strength of the technology. Figure 28 shows another solar roof where the solar cells — thin-film amorphous silicon cells — have been integrated into the glass roof tile itself, giving a very pleasing visual effect in this case.

A million more

The Japanese initiative in embracing residential photovoltaics on a large scale prompted responses in both Europe and the USA. The European Commission prepared a Green Paper on photovoltaics in November 1996, resulting in a White Paper accepted in November 1997. This called for one million solar residential systems before the year 2010, with 500 000 in Europe and 500 000 in the developing world, to be subsidised by the Commission. In June 1997, President Clinton announced a similar one million roof target in the USA. Since then, several other countries, including Germany, Italy and the Netherlands, have announced their own targets for residential photovoltaics. The most recent addition to this list has been Australia where a US$20 million program over four years has been announced. This should allow the installation of about 10 000 rooftop systems.

Figure 28
Photograph of glass-roof titles with thin-film photovoltaic cells built into the tile. (Photo courtesy of the Sanyo Corporation)

Most of these programs involve government subsidies in one form or another, often as a rebate in the 20–50 per cent range. In essence, a three-way subsidy is often involved with subsidisation by the government, the home-owner and the participating utility. The subsidy from the government generally takes the form of a rebate on the purchase price. The home-owner also subsidises the system because it generally still would be cheaper to buy electricity from the grid, even with the government rebate. The grid also gets involved by providing 'net-metering', whereby it buys electricity back from the home-owner at the same price it charges for electricity going in the other direction. In this way, the utility provides the storage role previously noted, without charge (as well as losing revenue from electricity it would sell to the household if the solar system was not there).

The diversity and range of these international initiatives has created a high level of confidence within the photovoltaic industry that there will be rapidly expanding markets for its products over the coming decade. This is encouraging investment in new manufacturing facilities and in the commercialisation of new technologies.

Sydney 2000 Olympic Village

The athletes' village for the Sydney 2000 Olympics is an example of the type of residential development that might be common in the future. Some of the 665 residences, built to become a legacy of this Village and to be sold on the open market before and after the Games, are shown in Figure 29. Each house was designed for low energy consumption by using efficient electrical appliances, gas for cooking, a gas boosted solar hot water heater, as well as having a 1-kilowatt photovoltaic system on its roof. Taking into account the low energy use, this system was sized to generate the average electricity requirements of the home.

Figure 29
Some of the solar-powered houses built to provide accommodation for the athletes participating in the Sydney Olympics.

An interesting point arises from this development. Although the homes were priced in the US$220 000–400 000 range, the cost of the photovoltaic system was about US$7000. At this price, the photovoltaic system does not represent a particularly attractive investment in terms of the value of the electricity it will generate, using standard economics. However, it adds only an additional 2–3 per cent to the price of these homes, which makes the systems 'affordable' if financed as part of the housing package.

AC modules

At present, it takes a trained electrician to connect the photovoltaic system to the grid network. The DC outputs of the modules are combined and fed into a single inverter whose output is connected to the grid.

All this could become a lot simpler. Recently, small inverters have been developed that are attached to the rear of individual modules as in Figure 30. The DC output of the module is converted straight into AC at the module level, producing an 'AC module'.

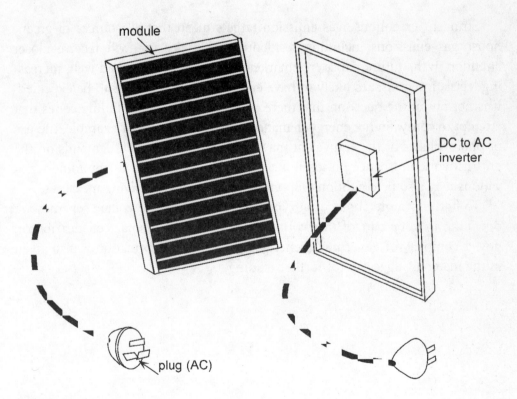

Figure 30

AC module concept. The dedicated inverter converts the DC output of the module to AC, the common form of electricity in city housing.

AC modules have a number of advantages. They greatly simplify the connection of the modules together and also reduce system losses. As suggested by Figure 30, for small systems, an additional benefit is that it may soon be possible just to plug them into power outlets, like a normal household appliance — this appliance generates electricity rather than consuming it, however. (The inverter is designed to operate only when connected to the household power supply so there is no electricity at the plug unless it is plugged in. If the photovoltaic system is too large, however, there is the chance, with this approach, of overloading the household wiring if a fault develops in other household equipment.)

A plug-in photovoltaic appliance available at the local supermarket or hardware store for those wanting to reduce their electricity bill and the amount of carbon dioxide being pumped into the atmosphere may be just around the corner in some countries. Until then, the local solar cell distributor or utility company is the best point of contact for those wanting to 'solarise' their homes.

***

Stringent greenhouse gas emission targets mean that all sources of greenhouse gas emissions including residential electricity use will receive closer attention in the future. Since free market forces have not worked well, increasing legislation on issues likely to have environmental impact is to be expected. It is not out of the question that there could eventually be building codes that attempt to constrain the energy demands of new housing. For example, the use of photovoltaics or the equivalent may be stipulated to lessen demands on the grid network and hence on fossil fuel emissions. Approvals for building renovations may also be conditional upon taking such energy-saving measures.

To find out more about what types of residential photovoltaic schemes are operating in your part of the world, contact your local solar cell distributor, power company or government energy agency. More information is also given in the material suggested for further reading.

Further reading

Erge, T. et al. (1998), 'The German 1000-roofs-PV programme — a resume of the 5 years pioneer project for small grid-connected PV systems', *Second World Conference and Exhibition on Photovoltaic Solar Energy Conversion*, July, Vienna, pp. 2648–51. (Gives a good summary of lessons learned from the German '1000 Roof' program as well as further references to more detailed information.)

Honda, T. (1998), 'NEDO's solar energy program', *Renewable Energy*, vol. 15, pp. 114–18. (Gives a brief outline of the Japanese photovoltaic program.)

Strong, S. (1991), *The Solar Electric House*, Chelsea Green, Vermont. (Written by one of the pioneers in the residential use of photovoltaics, this book anticipates many of the recent developments in this field.)

7

ARCHITECTURAL PHOTOVOLTAICS

Glamour and power

Another way of offsetting the present high cost of solar cells is to add value additional to that from the power they produce. At present prices of US$260–570 per square metre of module area (roughly double this for installed costs), photovoltaics are well below the price of premium building cladding such as polished marble. If the architectural value of this new medium is accepted, the electricity produced from it becomes a bonus.

In an architectural sense, photovoltaics can make a strong statement about a building and its occupants: modern, high-tech, socially and environmentally responsive. Photovoltaics can turn what otherwise would have been a mundane building project into one attracting local and international attention.

Although this use of photovoltaics is in its infancy, companies (mainly in Europe) now specialise in photovoltaic product for integration into buildings. There is also sufficient past experience to ensure the technical success of such installations.

How much power can be generated from coating buildings in this way? Even in the United Kingdom, not noted for its sunny climate, coating all suitable building surfaces with photovoltaics would give an electricity-generating capacity exceeding that of the country's conventional generators. The situation could only be better in most other countries of the world. Figure 31 shows an example of the type of system involved: here the Doxford Solar Office makes good use of a photovoltaic curtain wall. This building has been orientated so that the solar panels point southerly, on a slope to increase solar output.

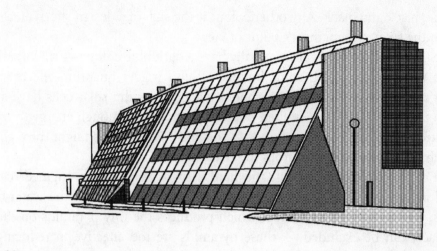

Figure 31
Solar Office of Doxford International.

The right direction

How important is it to try to have the photovoltaics pointed in the right direction? The answer, for Europe at least, is that the highest annual output is obtained if the system points a little west of south at a tilt of 30–40 degrees to the horizontal. The good news is that the output remains high even if the solar system is pointed well away from this direction.

For example, a vertical system pointing due south will still give 70 per cent of the best possible output. It performs well when the sun is low in the sky (mornings, evenings and most of winter), but less well in summer. A vertical system pointing due east or west gives 50–60 per cent of the maximum output, doing well in the morning and evening, respectively.

A vertical wall facing the north in Europe never sees direct sunlight except for short periods in summer, early in the morning and late in the evening. However, it still gives 20–30 per cent of the best possible output due to diffuse light scattered across the sky, even on clear days, but especially in cloudy weather.

Orientating the system so that it points vaguely towards the sun is a good idea, but not essential for reasonable outputs. Often electricity is in most demand on summer afternoons due to building air-conditioning loads. A system pointing more westerly than easterly could be the best choice in this case.

Choice of colour

Colour becomes important once aesthetics are involved. For best performance, solar cells would have the same colour choice as the original T-model Fords —

black, black and black. Any other colour is the sign of a design shortfall, at least from the solar cell engineer's point of view.

Silicon wafer-based cells normally have a dark blue colour — silicon reflects blue photons very well. Some bounce off and trigger quantum effects in our eyes instead of in the cell. Amorphous silicon thin-film solar cells have a reddish tinge — the low-energy red photons do not have enough energy to release electrons from the chemical bonds in this material and go straight through, but reach the eye by reflection from the rear contacts.

For the architect's benefit, the hard-earned gains of the engineer can be undone to give a range of colour choices. For example, with the wafer-based cells, the surface texturing step, which produces the tiny pyramids on the cell surface, can be excluded — these pyramids are too effective in reducing the number of photons reflected, hence preventing the appearance of strong colours. Once the pyramids are removed, the colour of the photons reflected can be controlled by adjusting the thickness of the thin interference film placed along the top cell surface.

Most work in this area has been done by BP Solarex. The top-of-the-range buried contact cells (see Chapter 4) this company markets (as the 'Saturn' line) can match the performance of standard cells even when sacrifices are made to get the right colour. The company has developed modules in four different colours — the standard 'dark blue', light blue ('steel blue'), purplish ('magenta') and yellowish ('gold'). These coloured modules have been used only in a limited number of applications to date, however, and would be expected to attract a premium price.

Transparent cells

Silicon wafers are opaque. By packaging wafer-based cells between glass sheets with plenty of space between the cells, semi-transparency can be obtained. Quite interesting internal architectural effects are possible as shown in Figure 32.

More uniform transparency can be obtained with 'thin-film' cells in two ways. If metal contacts are replaced by 'transparent conductor' layers (Chapter 3) and the cells made very thin, some of the photons will pass right through the cell (see Figure 19, page 41). This approach is particularly suited to amorphous silicon modules. The other approach is to pattern a large number of tiny holes into the thin-film material. If these holes are small and close together, the eye will just see a transparent layer, with its transparency determined by the fraction of thin-film material removed.

A completely different approach to transparency is possible with the dye cells that mimic photosynthesis, briefly described in Chapter 3. Unlike

Figure 32
Semi-transparent glass facade
with integrated photovoltaic
modules on the office building of
HASTRA Electricity Works,
Hannover. (Photo courtesy of
HASTRA, Germany)

semiconductors, the dye molecules in this cell absorb only a narrow band of light colours — normally a disadvantage when it comes to getting high performance. However, if this band were at infrared wavelengths, the cell would let all the visible photons through. The module could look as clear as normal glass. (Performance would suffer — a silicon wafer cell using only the infrared photons in sunlight would give only one-third of its normal power.)

Innovative ideas

Since the technology is so new, photovoltaics allow plenty of scope for architectural creativity. International competitions encouraging new ideas in their use are held every few years and help explore the architectural potential of this new medium. A common use is to replace glass with solar cells as structural glazing in facades and curtain wall applications. One elegant example is shown in Figure 33.

A more vigorous use of photovoltaics is shown in Figure 34. Here the photovoltaic modules not only generate electricity but also act as shades, as well as collecting heat for the building.

Figure 33
Award winning solar facade application of
thin-film photovoltaics (Flachglas Building,
Wemberg)

Figure 34
Photovoltaic facade providing power, heat
and shade. (Photo courtesy of Atlantis
Energy Ltd)

A different approach has been taken in the roof-integrated solar system on the headquarters of Digital Equipment in Geneva (Figure 35), creating a modern, high-tech image appropriate to the company's business.

Figure 35
Headquarters building of Digital Equipment, Geneva. (Photo courtesy of Atlantis Energy Ltd)

A final example of an adventurous use of photovoltaics is to power the Olympic Boulevard Lighting Towers for the Sydney 2000 Olympics (Figure 36). The towers add a sense of fun and excitement to this large boulevard, providing a balance between the massive scale of the facilities lining it and that of the people using it. They also serve as meeting places for the latter. The photovoltaics, mounted on the horizontal support elements near the tower base, provide shade as well as power for the night-lighting and electronic information screens.

Figure 36
Sydney 2000 Olympics Boulevard light towers each powered by the photovoltaic shading platform near the tower base. (Photo by Juliet Byrnes, SOLARCH, UNSW)

Small is beautiful

The cost of architectural photovoltaics, particularly for smaller systems is expected to be significantly reduced by the use of small reliable, cheap inverters. As previously mentioned, the DC outputs of solar modules are usually connected and fed into a single inverter, which converts the combined DC output into AC. These inverters take advantage of the economy of size — big inverters are cheaper per unit of power converted than small inverters. Or are they? There is also an economy of volume — the more of something that is made the cheaper it usually becomes, as already seen for the solar cells themselves .

Since AC wiring is by far the most common wiring in buildings, another approach starting to be used is to directly convert the DC electricity to AC at the output of each module — the 'AC module' concept of Figure 30 (page 61). Many more of these module-sized inverters are required, pushing the costs down compared to larger units. Costs of connecting the systems together are also reduced. Some operational features, such as tolerance to shadows across the system, are also improved.

***

Photovoltaics are well suited for architectural use since they can make a distinctive statement about the building involved. Costs are already within the range of standard building cladding material.

Pioneering installations have removed most of the engineering challenge from this application. What is now needed is for innovative architects to adopt this new design medium, even as a trademark. No longer are the inefficient buildings of the previous few decades good enough — buildings of the future can be environmentally responsible as well.

Further reading

Sick, F. and Erge, T. (1996), *Photovoltaics in Buildings: A Design Handbook for Architects and Engineers*, James and James, London. (This International Energy Agency handbook 'will enable all building architects, engineers and property owners to make the integration of photovoltaics into buildings an architecturally appealing and energetically effective option'; see also Humm, O. and Toggweiler, P. (1994), *Photovoltaics and Architecture*, Birkhäuser.)

Various papers (1996), 'The future of photovoltaics in the built environment', special issue of *Progress in Photovoltaics*, July–August, Wiley. (A collection of papers on the architectural use of photovoltaics from a range of perspectives.)

8

ENERGY FOR THE DEVELOPING WORLD

One billion people in the developing world still lack access to
clean water ... nearly 2 billion lack adequate sanitation
... electric power has yet to reach 2 billion people

World Bank Report, 1994

Power for the world

Over half of the six billion or more people on this planet live in rural areas,
mainly in the so-called developing countries — more specifically, countries with
low per-capita income (as can be calculated by dividing the gross national product by the population count).

Energy is probably not seen as the most pressing problem in these countries.
More immediate concerns include poverty alleviation, food security, access to
health, education and employment, rural-to-urban migration, international
trade, foreign exchange and, often, the presence of civil war. The main energy
problem in rural areas is usually finding enough wood for cooking meals. In
some countries, fuel for cooking accounts for more than half the national energy consumption.

Most developing countries are in the process of steadily extending their electricity network into rural areas in an attempt to improve the quality of rural life.
This approach is too expensive to be a complete solution since the cost of the
distribution lines typically runs at US$20 000–30 000 per kilometre. Such costs
can be recovered only in areas of high population density and where customers

can afford to use plenty of energy. Although this form of rural electrification should and will continue, it will not meet the needs of large segments of the population in the near or medium term.

Photovoltaics can provide a complementary approach to such electrification programs. Their modularity allows photovoltaics to generate electricity right where it is needed. As seen earlier, subsidised programs in the richest parts of the world, encouraging the urban residential use of photovoltaics, are helping to drive down photovoltaic prices. This downwards spiral will allow photovoltaics to better meet the more modest power needs in some of the world's poorest rural areas.

In his paper, 'Power for the World', Wolfgang Palz of the European Commission in Brussels calls for a massive 20-year international initiative to supply electricity, using photovoltaics, to over one billion people worldwide. Palz calculates that about 10 watts of photovoltaics per person is needed for 'survival needs' and the 'development and basic needs' shown in Table 1. He calculates the investment required as US$3 billion per year over the 20-year period (Palz points out this is 3 per cent of yearly over-all energy investments in developing countries and less than 0.5 per cent of global military expenses).

How feasible is this target and what are the main issues? There is already plenty of experience with the type of systems Palz describes, with over 200 000 'solar home systems' already operating worldwide, notably in China, Colombia, India, Indonesia, Kenya, Morocco and Mexico. There are also over 10 000 solar-powered water pumps providing drinking water and water for irrigation. Solar-powered outdoor lighting is also widely used, such as for adult literacy classes in India or village meeting places in Indonesia. Photovoltaics are also widely used for rural health centres for tasks such as vaccine refrigeration (domestic refrigeration may not be widely used in some communities in developing countries whose traditional foods do not require cold storage).

It has been proposed that the rate of adoption of new technology depends on its 'relative advantage' over that it supersedes (or over other possible uses of available funds). The degree of relative advantage might reflect economic or social issues. The access to radio and television that a solar home system provides could become an important driver for families to make the economic sacrifice required to acquire, or even maintain, such systems. The likely development of low power television sets using liquid crystal as in portable computers should boost this market. Schemes involving individual ownership and responsibility for the solar system and some type of financial sacrifice on the part of the user, even if nominal in terms of the actual system costs, have generally fared better than those involving larger community-based systems.

Table 1
Photovoltaics for minimum needs (scheme for a model village of 600 people in 40 families of 15 each, by Wolfgang Palz).

	Watts/Capita
Survival needs	
Water disinfection and supply	3.15
Health centre	0.35
Emergency telephone	0.05
Dental care	0.45
Total:	4.00
Development and basic needs	
Family lighting	0.87
Street light	0.40
Cultural centre	0.33
Insect fighting	0.50
Educational TV	0.23
Battery charging station	0.34
Rural telephone	0.03
Refrigerator for village shop	0.17
Workshop	0.68
Total:	3.50
Technical losses (battery …)	2.50
TOTAL	**10**

Solar home systems

A typical individual solar home system is shown in Figure 37. Unlike its grid-connected urban counterpart, the DC output of the photovoltaics is used directly rather than being converted to AC form. The photovoltaic module is connected to a lead-acid battery and the household electrical loads through an electronic 'charge-controller'. This controller prevents the battery from overcharging during sunny weather and also prevents it from going flat, caused by drawing too much electricity from it. Without this controller, battery life would be drastically reduced as explained in Chapter 5.

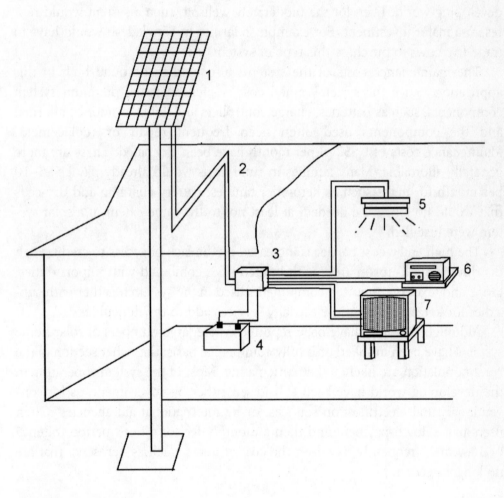

Figure 37
Solar home system for rural electrification: (1) solar module (in this case pole-mounted, ensuring access to sunlight and making unauthorised removal difficult); (2) wiring to house: (3) charge controller; (4) battery; (5) fluorescent lamp; (6) radio; (7) television.

The solar panel generally would be 20 watts or 50 watts rating. The smaller size could power lighting and a radio while the larger system could also run a black and white television set. Lights would generally be efficient fluorescent lamps or very low power incandescent lamps, used to provide night 'orientation'. Using the solar system to power the radio removes the need for dry-cell batteries, which can consume a significant part of the annual earnings of a rural family. Other major expenses displaced by the solar system are those of kerosene, candles or paraffin, used for lighting.

How affordable are such systems? The answer is that even such a small system, costing US$500–800, is beyond the resources of most in rural areas of the developing world. Even for the moderately well-off, such a system would represent a major investment. For example, a farmer in Bangladesh would have to trade five cows to purchase this type of system.

The maintenance costs of the systems also tend to be quite high in this application. Since the system capital cost is such an issue, premium system components, such as batteries, charge controllers and lamps cannot be afforded and the components used often need frequent repair or replacement. Maintenance costs of US$2–3 per month have been estimated. These are more generally affordable. Many families in rural areas would already pay US$5–10 per month for items such as kerosene, candles, battery charging and dry-cells that would no longer be needed, at least not to the same extent, if a solar system were installed.

The high entry-cost barrier is increased by the lack of access to credit. High distributor mark-ups in areas with limited sales combined with import duties, tariffs and taxes on imported components used in the system together with subsidies for kerosene and grid-electricity services add to the difficulties.

Although examples have been reported where large numbers of solar home systems have been installed on a fully commercial basis, the richer section of the rural population are likely to be participating. Most of the systems operating in the developing world have been subsidised either by government, as for conventional rural electrification services, or by international aid agencies. Often users pay a down-payment and then a monthly fee for a fixed period (often 5 or 10 years), frequently based on the cost of items, such as kerosene, that are no longer needed.

Revolving funds

The 'revolving' fund approach increases the benefits from a starting sum allocated for such systems, although experience with such schemes has not always lived up to expectations.

Erik Lysen, formerly of the Netherlands' Agency for Energy and the Environment, discusses the following example. Say we start with a US$50 000 fund, from which 100 solar home systems at US$500 can be purchased. At 10 per cent interest and a 10-year repayment period, the required monthly repayment would be $6.78 or $81.37 per year. During the first year, $8137 would become available for the purchase of an additional 16 systems. During the second year, an additional 18 systems could be purchased, and so on. After 10 years there would be over 280 systems operating, with no further payments required on 100 of these. The number of systems on which payments are made drops back a little in Year 11 but continues to grow past that point. Lysen calculates that 1672 systems would be operating in year 20. If the scheme were then terminated, payments coming in over the next 10 years would give an accumulated capital fund of $689 000 corresponding to 9.1 per cent compounded interest on the original $50 000.

This is perhaps not too surprising since the scheme is producing 10 per cent return on funds borrowed (not all funds are generating this return in the final 10 years, in this example). This is an ideal example since it assumes no operating costs for the scheme, no maintenance costs for the system, and no default on payments. The latter has often been a problem in practice, particularly when there has been a technical problem with the system or if it has not otherwise lived up to expectations (the 'relative advantage' has to be real, particularly with such a large investment).

This means that, in some cases, the funds 'did not revolve'. Lysen calculates, with a 5 per cent drop out rate (apparently without repossession of the system and its resale), the number of systems installed over 20 years would drop from 1672 to 948 and the final accumulated funds to a more modest $88 000.

The key point with revolving fund schemes is that repayments are used to increase the number of systems beyond those financed by the initial fund. If these funds came as a grant, the repayments on the system could be based on what could be afforded in the region involved, rather than on normal economic returns. The repayments could still be invested in additional systems. For example, in the previous example, if recipients were asked to pay $3 per month and owned the system after 10 years, ideally 233 systems would be operating after 20 years with sufficient funds coming in to finance about 4 new systems per year for well into the future (more would be possible if the costs of the photovoltaics systems decreased over this period).

Appropriate maintenance infrastructure, to ensure the systems can be kept operating, and incentives for non-default on payments are essential to the success of such schemes.

Technical issues

Although the photovoltaic panels seldom pose a problem, other parts of the system are often not as reliable.

Batteries are a problem. Normal car batteries, the most likely to be available locally at a reasonable price, are not ideal for photovoltaics. Only a fraction of the total charge stored in the battery can be drawn out without reducing battery life, already only a year or two under ideal conditions.

Car batteries require plenty of topping-up with distilled water and a lot of the stored charge is lost by internal self-discharge. Heavy-duty batteries for trucks and buses have the same problems, but to a lesser degree, and are a better choice. Traction batteries, such as those used for electric forklifts, are less likely to be available but are also a better choice since more of the stored charge can be used. Sealed lead-acid batteries are also a better option since, as well as this advantage, they require little maintenance.

Special batteries have been developed specifically for solar systems and, although more expensive and less likely to be available locally, these are most likely to perform well in this application. An operating life of up to five years is often quoted. Nickel-cadmium batteries are also suited for this type of solar application but are prohibitively expensive.

Another important system component is the charge controller, since this protects the battery and extends its life. Ideally, the controller prevents the battery from overcharging as well as from being too fully discharged. The controller can disconnect loads when the battery charge drops below a set level, a feature not often appreciated by the system user. Indicators, so users can readily 'see' how much charge is in their batteries and ration its use, are a better idea.

The reliability of the charge controller electronics has been an issue. These units are suitable for local manufacture, although high reliability is not likely without a large amount of development. The additional cost of this unit and uncertainty about its reliability sometimes lead to it being excluded, reducing likely battery life. In the field, users often bypass the unit if it malfunctions or interferes with the user's desired use of the system.

Even though solar modules are extremely reliable and need little attention, ongoing maintenance of the overall system is required, as is some source of advice for the user. Existing technically orientated infrastructure can be used. In China, for example, outlets for agricultural machinery now act as distribution and servicing points for solar home systems.

Training of local young people as technicians in this area, enabling them to earn income from this work, has also been recommended.

Supporting solar power

Experience to date has shown the importance of having the local community fully behind any initiative to install solar home systems. High levels of perceived relative advantage are important, such as the attraction of television. The attitude of regional electricity companies to the solar home systems is also relevant. Solar programs have worked best when conducted in cooperation with the rural-electrification programs of these companies, rather than in direct competition.

Some success stories

Although 95 per cent of Mexico's population is connected to the electricity grid, several million Mexicans live in remote rural areas where this grid is unlikely to penetrate. About 20 000 solar systems have been installed by private companies. Additionally, since 1991, the Mexican government has operated a photovoltaic rural electrification program in parallel with its grid extension program. The program is community focused, in that requests for solarisation must come from individual communities.

Preference is given to small communities (100 or fewer people) with average distances between houses of 50 metres or more, no prospects for grid-electrification in the next three to five years and no plans for large electrical loads such as motors. The community also has to be willing to contribute to system costs and to create a local organisation to support the project, during and after installation of the solar home system, and to collect fees from the community for maintenance and system expansion. About 85 000 communities in Mexico could meet these criteria.

A typical system, at least as supplied by this program, would consist of a 50-watt module, a charge controller, a battery and three fluorescent lamps of 13 watt rating (when grid-electricity is connected, lighting is the most common load since households often cannot afford other appliances).

Over 35 000 solar systems have been installed under the Mexican program, reaching over 700 000 people in more than 1000 separate communities. Although the scheme has not been completely without problems, it has established photovoltaics as an acceptable rural-electrification route within the country. Technical problems mainly related to the uneven level of reliability of the components of the system, with the usual problems with batteries, charge controllers and even fluorescent lamps. On the social side, fewest problems emerged when users were involved in the project as early as possible and properly trained in the use and maintenance of the system. On the institutional side, problems often related to the unfamiliarity of staff from the executing agency with the new technology, the absence of codes and standards for products, the lack of a local support industry and the immaturity of schemes for project implementation.

Indonesia is in the process of providing one million of its rural households with solar home systems over the next 10 years. A pilot project began in the village of Sukatani in 1989 followed by one in Lebak in 1990. When the Indonesian president first saw these systems, he was sufficiently impressed to launch a Presidential Aid program in 1991 that saw 3500 additional systems installed.

This led to an AusAID program, involving the Australian and Indonesian governments, to install a further 36 400 systems. Like the Presidential Aid program, this program is based on village cooperatives, which act as an intermediary between the government and the users. A down-payment equivalent to US$20 is made, with monthly payments of US$4 over 10 years. Of this, 10 per cent is a management fee to the cooperative, 20 per cent goes into a fund for battery replacement and 70 per cent is used in a revolving fund, financing other systems. A scheme involving an additional 35 000 systems has been recently announced involving the Bavarian and Indonesian governments.

An additional 200 000 homes in Indonesia are targeted over the next four years in a market-orientated approach supported by the World Bank. Systems are sold based on a hire-purchase contract between the householder and local dealers, responsible for system installation and maintenance. A down payment of between US$80–125 typically would be involved with monthly payments of US$8–10 over three to four years. No ongoing subsidies are involved, only an initial subsidy of US$75–125 on the cost of the system, depending on the location.

Village power systems

Most experience to date suggests the individual solar home system is a better approach to providing household power in rural villages than the alternative: a larger system shared by the whole community.

A number of village-level photovoltaic systems have been installed around the world that mimic the conventional grid supply in their operation. The system generally consists of a single centralised photovoltaic generator, a battery storage bank, a DC to AC inverter and a local electricity distribution network, carrying the electricity to the houses in the village. Typically, peak power is in the 1–20 kilowatts range.

There have been some problems with these systems, including a lack of reliability, particularly of the inverters. Another major issue is the management of such systems. Approaches, such as imposing a limit upon the daily energy consumption of individual users of a shared system, may well improve the future effectiveness of such village-based systems.

Water pumping

In many parts of the world, the role of collecting household water falls to women and children, who often walk several kilometres to reach a reliable supply. Consumption of unhealthy water is one of the major causes of disease worldwide.

Water pumping is well suited to photovoltaics since storage batteries are not required — pumped water can be stored instead of electricity. This makes the system simpler and, in principle, much more reliable.

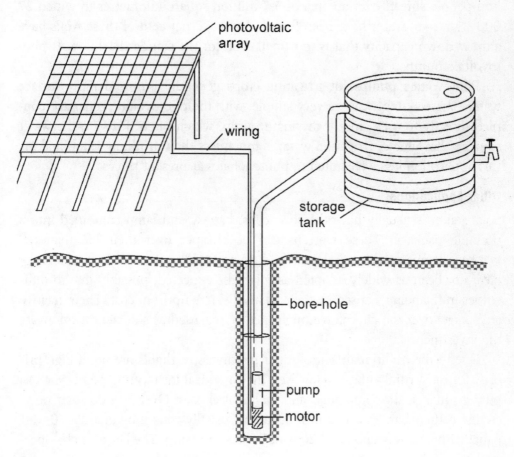

Figure 38
Photovoltaic-powered drinking water supply system (an electronically commutated DC motor is used in this example; an inverted would be required for an AC motor).

Photovoltaic pumps are particularly suited to regions where water is scarce because they are small and, as noted, they can be very reliable. They generally tap sources of water from twenty metres or so underground (Figure 38). Convenience of supply provides the relative advantage and is a key issue in the success of such projects.

Most drinking water systems of this type use 1 kilowatt or so of photovoltaics, beyond the resources of private citizens. Generally, systems are funded under international aid programs, with the recipient community responsible for system maintenance through sales of drinking water.

Wells often have a limited 'draw-down' capacity (the rate that water can be withdrawn without running the well dry). This means that a small photovoltaic system can be a good match for the maximum draw-down rate. For example, the countries of the Sahel (in West Africa) are home to some 35 million people spread over an area of 5.7 million square kilometres in which 27 000 sources of water have been found. About 70 per cent of these wells have a draw-down capacity that is too small even for the standard 1-kilowatt photovoltaic pump.

Photovoltaic pumps for irrigating crops generally pump shallow surface water. These systems can be very simple, with the photovoltaic generator coupled directly to a DC motor driving a pump, which often floats on the water source. The value of irrigation water is much less than drinking water, making this application less well suited to photovoltaics at present prices.

Other applications

Solar lanterns usually have the solar cells, battery and lamp combined into a portable package. These lanterns can be cheaper over their life than the kerosene lamps used in rural areas and do not smell or present fire risks. They have not been as widely adopted as might be expected, possibly due to difficulties in financing, caused by their relatively large up-front cost. Their 'relative advantage' over the alternative might not be regarded as sufficient to motivate the investment.

Due to the traditional lack of night lighting, streetlights are not a high priority for most rural villages. However, they are ideal for meeting points such as public squares, shopping areas and for night classes. Over 30 000 solar powered streetlights are installed in India alone, usually consisting of a 25–70 watt panel with a battery and a 10–20 watt fluorescent lamp. The lamp can be operated several hours per night.

Solar-powered telephones usually are coupled to the standard telephone network using a radio link. The use of telephones is becoming more common

as rural populations increase their contact with cities, often through the migration of family members.

Solar power can improve productivity in making handicrafts by providing
electricity for lighting and for powering small tools, such as drills and saws.

In some parts of the world, a common way of getting electricity to homes
not serviced by the grid is to use a lead-acid battery, often a car battery, as the
electrical equivalent of bottled gas. After discharging, the battery is taken to a
nearby re-charging station, normally twice a month. A number of solar-powered
battery charge stations have been established at different places around the
world.

Rural health centres require electricity for uses such as light, water pumping, sterilisation, refrigeration of vaccines and so on. Photovoltaic systems in the
300–1500 watt range are often used to provide power for such needs.

Schools are an obvious target for community-based photovoltaic systems.
The photovoltaic power can be used to provide drinking water as well as elec-
tricity for powering educational television and lighting for adult night classes.
Typically 1 kilowatt of photovoltaic power would be used.

***

Although there are large numbers of photovoltaic solar home systems
already operating in the developing world, there is scope for increasing their use
several thousand-fold. The ongoing decrease in photovoltaic costs will accelerate the spread of the technology. A major challenge, however, is to develop the
infrastructure needed to allow the financing, installation and maintenance of
these systems. Major programs around the world are consolidating experience
in these areas and consensus on the best approaches is starting to form.

Further reading

Lorenzo, E. (1997), 'Photovoltaic rural electrification', *Progress in Photovoltaics,* vol. 5, January-February, Wiley, pp. 3–28. (Also see the special issue of this journal, 'The future of rural electrification', vol. 6, September–October, 1998, containing several papers on this topic.)

Ossenbrink, H. (ed.), (1998), *Proceedings of the Second World Conference on Solar Energy Conversion,* Vienna, July, pp. 2859–3039. (These pages of this conference proceedings contains papers on 'Stand-alone systems and their applications'; 'Rural electrification'.)

Palz, W. (1994), 'Power for the world', *Proceedings of the 12th European Photovoltaic Solar Energy Conference,* Amsterdam, April, pp. 2086–88. (This conference proceedings contains the papers from the symposium 'PV in developing countries', pp. 1923–2088, including Uken, E. 'Affordable light for unelectrified dwellings', pp. 2012–15, and a keynote paper by Lysen, E. 'Photovoltaics in the south', on which some of the material in this chapter was based.)

Roberts, S. (1991), *Solar Electricity: A Practical Guide to Designing and Installing Small Solar Systems,* Prentice Hall, New York. (As the title suggests, this book is a good 'hands-on' source for those involved with installing systems in the developing world.)

9

POWER FOR THE FUTURE

Times are changing. The 20th century spanned a period of massive growth and development powered by solar energy congealed in fossil fuels. This approach to energy supply can only be a 'temporary fix' because of its damaging effects on the environment from carbon dioxide emissions.

The sun already does 99.99 per cent of the work in supporting human life as we know it. The energy reaching the earth from the sun in only three weeks equals that stored in all known reserves of fossil fuels. There is more than enough energy coming from the sun to replace the relatively trifling amount now contributed by fossil fuels.

Solar cells provide the most attractive way yet suggested of tapping into this massive resource. Large centralised power stations, like the fossil-fuelled plants now used to generate most of our electricity, eventually will be feasible with photovoltaics. However, as discussed, the technology can also be used on an individual basis, most notably on the roofs of private homes.

We have seen how solar cells almost magically convert photons of sunlight into electricity. Improved technology, such as that offered by 'thin-film' approaches, will contribute to a continuing downward spiral in cell costs. Eventually, the cells will cost only a few times the cost of glass sheeting. Over the coming decades, photovoltaics will move from its past position of being an expensive way of generating electricity to being one of the cheapest.

Fortunately, demand is escalating, with many countries now committed to the large-scale use of photovoltaics on the roofs of private residences in areas already serviced by conventional electricity. The tens of thousands of homes powered by photovoltaics at present will grow to millions over the coming decade, under current programs. Within a decade, driven by these programs, cell costs are expected to be low enough for these household systems to be competitive without subsidies.

Architectural use in buildings is also likely to become more significant over the next 10 years. The amount of energy that can be supplied in this way is significant. If all suitable facades and roofs in urban areas were coated by photovoltaics, their combined capacity would be comparable to that of all the standard fossil-fuelled generators in most countries, even in the less than ideal European climates.

These applications in the wealthier countries will accelerate the spread of another important use — supplying power to the two billion people worldwide presently without it. Photovoltaics is already the cheapest way of doing this, particularly in the remoter and less accessible regions of the developing world. Although hundreds of thousands of photovoltaic solar home systems are already installed in some of these regions, decreasing photovoltaics prices will reduce the major impediment to their more widespread adoption — the difficulty of financing. The total number of systems would need to grow a thousandfold to make a significant impact on needs.

Although the three applications above — household power in both the developed and developing world and use on other buildings — are thought likely to be the most important over the coming decade, they are not the limit on what is possible. There is no technical reason why photovoltaics cannot be used to supply the vast majority of the world's energy needs, in the longer term. The key requirement is a reduction in cost to a few times that of standard glass sheeting, a target that is technically feasible. The other is some improved means of energy storage or transmission, such as storage as a chemical fuel or a low cost electricity transport approach. From about 2050 onwards, this 'human-scaled' technology could be the world's dominant source of energy.

GLOSSARY

AC (alternating current)
Electricity, as in normal homes, where the electrical current changes direction 50 or 60 times per second, depending upon the country where you live.

charge controller
A box of electronics which prevents batteries from being overcharged.

DC (direct current)
Electricity, as supplied by chemical batteries and solar cells, where the electrical current flows in the one direction (from positive to negative terminals).

doping
The process where small quantities of foreign atoms are introduced to control the electronic properties of semiconductors.

EFG (edge-defined film-fed growth)
A method of forming silicon in the form of a thin solid ribbon by drawing it through a carbon die from the melt.

electron
The fundamental carrier of electric charge. Electrons in motion represent electrical current. Electrons carry negative charge.

hole
An effective electrical charge carrier formed by the absence of an electron, as a bubble is formed by an absence of liquid. Holes carry positive charge, equal but opposite to that of an electron.

inverter
A box of electronics that converts the DC output of a solar module to the AC form required by common household appliances.

load
Any appliance or device that can be powered by electricity.

module
A self-contained package of solar cells, generally consisting of a glass sheet covering 36 cells connected together and often with an aluminium frame around its edge.

photon
A fundamental quantum of energy into which light can be subdivided.

p-n junction
An electronic device formed by the interface between a region of semiconductor doped to have 'positive' (p-type) properties and a second region doped to have 'negative' (n-type) properties.

quantum
The smallest package of energy that can be exchanged in any physical process.

RAPS (remote area power supply)
A self-contained photovoltaic system designed for remote areas where there is no auxiliary source of electricity.

renewables
Energy sources such as wind and solar that are constantly renewed and not depleted during use.

substrate
A supporting layer on top of which other material can be deposited.

superstrate
A supporting layer, transparent for solar use, on the bottom of which other material can be deposited.

transformer
A commonly used piece of electrical equipment that changes the voltage of AC electricity (for example, a transformer changes the high voltages used for distributing electricity to the lower voltages used by the home).

USEFUL CONTACTS

The World Wide Web is an excellent source of information on photovoltaics (PV) which is regularly updated. The following list of solar cell web sites is based on the list compiled by Dr Mary Archerincludes (thanks also to Pietro Altermatt, Holger Neuhaus and Johnny Wu). For information on what is happening in your area, check the web site of your local power company or energy agency.

www.caddet-re.org/html/pvpsp.htm
    (International Energy Agency's PV Power Systems Program)
www.energy.sourceguides.com/businesses/byP/solar/pvM/byN/byNameB.shtml
    (PV businesses in the world by name, with links to manufacturers' web sites)
www.epa.gov/globalwarming/actions/solar/sol/index.html
    (US Environmental Protection Agency's web site, describing the benefits of using solar energy to mitigate global warming)
www.eren.doe.gov/consumerinfo/refbriefs/t396.html
    (List of web sites concerned with use of PV and other renewables in developing countries)
www.eren.doe.gov/pv
    (US Department of Energy's photovoltaic program)
www.eurosolar.org/Publications/100.000.roof.html
    (Germany's 100,000 Roof Photovoltaic Program)
www.fsec.ucf.edu/PVT/index.htm
    (The Florida Solar Energy Center's photovoltaics information homepage)
www.greenhouse.gov.au
    (Australian Greenhouse Office, including details of PV residential program)
www.iclei.org/efacts/photovol.htm
    (Provides easy-to-understand fact sheets on solar cell technology)
www.ise.fhg.de/Research/SWT
    (Solar Cells Division, Fraunhofer Institute for Solar Energy Systems, Freiburg, Germany)
www.nef.or.jp/english/index.htm
    (Provides information on the Japanese residential PV program)

www.nrel.gov/photovoltaics.html
> (PV activities of the US National Renewable Energy Laboratory)

www.pv.unsw.edu.au
> (Centre for Photovoltaic Engineering, University of New South Wales, with links to other PV sites)

www.pvpower.com
> (Jobs, training and events in PV)

www.seda.nsw.gov.au
> (Sustainable Energy Development Authority, Sydney, Australia with information on PV and Australian programs)

www.solarschools.com/uk/scolar.htm
> (The UK Scolar Program to install PV in schools and colleges)

INDEX

"Love & love again & again for hating again and again
loses the heart & gains a grudge against the beauty of
life."